Economic Evaluation of Sustainable Development

"This book focuses on the many problems that arise in evaluating past decisions and informing pending ones over the entire spectrum of policies relating to social and economic development. It takes readers on a guided tour of three main approaches, citing real-world examples in each case. Readers can gain much from the knowledge and wisdom reflected in its pages, based on the many decades its authors have spent in the struggle to improve human conditions around the world. Here one can learn not only about hopeful approaches, but also about their limitations and the many real-world obstacles that must be confronted in their application. Overall, a serious and honest attempt to inform readers of how complex are the problems of accelerating economic and social development, and a call for greater application of available techniques and approaches in this struggle."
—Arnold Harberger, Distinguished Professor Emeritus, *University of Chicago* and *University of California Los Angeles*

"The book's central theme—that evaluation should inform economic policy making—is a key message for those in the business of sustaining development, growth and poverty reduction."
—Emmanuel Jimenez, Executive Director, *International Initiative for Impact Evaluation*

"This book is an excellent primer on (i) economic approaches to evaluation, (ii) objectives-based evaluation and (iii) their application to the framework of the Sustainable Development Goals. It will prove essential to students and scholars, but also to policy analysts who need practical methods to evaluate projects and policies in the new terrain, and to understand the conceptual and analytical foundations of proposed methods."
—Ravi Kanbur, T.H. Lee Professor of World Affairs, *Cornell University*

"In a world awash with data, very few people know how to move from data to evidence. This book is an invaluable guide to the tools and techniques that can be used. It links the disciplines of economics and evaluation and shows how the big challenges of our times—like achievement of the Sustainable Development Goals—depend on us all learning how to learn."
—Homi Kharas, Interim Vice President and Director, *Global Economy and Development, The Brookings Institution*

"This book is important and timely. It provides a detailed road map towards excellence in the evaluation of sustainable development at the higher plane of global policy."
—Robert Picciotto, Former Director General, *Independent Evaluation Group, The World Bank*

"Thomas and Chindarkar do an excellent job of providing tools to practitioners and guide them through the application of economics tools to evaluations of sustainable development. Impact evaluation, cost-benefit analysis, and objectives-based evaluations are married in a very nice collection of chapters that teaches and advocates high quality work. I recommend it highly."
—Jyotsna Puri, Head, *Independent Evaluation Unit, Green Climate Fund*

"The strengths of this book are its broad approach to policy evaluation—embracing both impact evaluation and economic analysis—and the clarity of its exposition, backed up by many examples. This is a very welcome contribution."
—Martin Ravallion, Edmond D. Villani Chair in Economics, *Georgetown University, Washington DC*

"This book offers a timely and most needed contribution to the field of evaluation, by bridging the gap between economics and evaluation methods. Thomas and Chindarkar have succeeded in presenting a compelling discussion on the application of economics approaches to traditional evaluation practices to improve the assessment of policy, program and project development effectiveness."
—Marvin Taylor-Dormond, Director General of Evaluation of the Asian Development Bank

"This book has brought together and integrated major approaches to the evaluation of development policies and projects and indicated their strengths and complementarities in a much needed way. The reader will learn and benefit much from it."
—George S. Tolley, Professor Emeritus of Economics, *The University of Chicago*

Vinod Thomas • Namrata Chindarkar

Economic Evaluation of Sustainable Development

palgrave
macmillan

Vinod Thomas
Lee Kuan Yew School of Public Policy
National University of Singapore
Singapore, Singapore

Namrata Chindarkar
Lee Kuan Yew School of Public Policy
National University of Singapore
Singapore, Singapore

ISBN 978-981-13-6388-7   ISBN 978-981-13-6389-4 (eBook)
https://doi.org/10.1007/978-981-13-6389-4

Library of Congress Control Number: 2019934472
© The Editor(s) (if applicable) and The Author(s) 2019, corrected publication 2019
This book is an open access publication
Open Access This book is licensed under the terms of the Creative Commons Attribution 4.0 International License (http://creativecommons.org/licenses/by/4.0/), which permits use, sharing, adaptation, distribution and reproduction in any medium or format, as long as you give appropriate credit to the original author(s) and the source, provide a link to the Creative Commons licence and indicate if changes were made.

The images or other third party material in this book are included in the book's Creative Commons licence, unless indicated otherwise in a credit line to the material. If material is not included in the book's Creative Commons licence and your intended use is not permitted by statutory regulation or exceeds the permitted use, you will need to obtain permission directly from the copyright holder.

The use of general descriptive names, registered names, trademarks, service marks, etc. in this publication does not imply, even in the absence of a specific statement, that such names are exempt from the relevant protective laws and regulations and therefore free for general use.

The publisher, the authors and the editors are safe to assume that the advice and information in this book are believed to be true and accurate at the date of publication. Neither the publisher nor the authors or the editors give a warranty, express or implied, with respect to the material contained herein or for any errors or omissions that may have been made. The publisher remains neutral with regard to jurisdictional claims in published maps and institutional affiliations.

This Palgrave Macmillan imprint is published by the registered company Springer Nature Singapore Pte Ltd.
The registered company address is: 152 Beach Road, #21-01/04 Gateway East, Singapore 189721, Singapore

FOREWORD

The emergence of a range of economic, social, and environmental threats to progress has led to a growing focus worldwide on sustainable development. The crucial questions that Vinod Thomas and Namrata Chindarkar address in this book are: What policies and investments better ensure sustainability? How best can sustainability be evaluated, reported, and acted on?

The 17 Sustainable Development Goals, addressing key socioeconomic, environmental, and governance issues, provide a framework for thinking about sustainable development. They place central importance on the nature of economic growth in terms of social inclusion, environmental stewardship, and good governance. These goals, surrounding "wicked problems" too complex to be dealt with directly, have received short shrift in country policies, despite their growing importance.

The challenge is to apply sound economic methods to analyze these complicated concerns and apply rigorous evaluation tools to assess progress. The book commendably confronts this challenge by connecting a set of evaluation tools with economic analytics, uncovers what seems to work and what does not, and indicates what more needs to be known. The result is practical guidance on development evaluation to a wide range of readers, including practitioners who conduct evaluations, students who wish to learn about evaluation, and policy-makers who make use of evaluation studies for decision-making.

Issues of sustainable development have a special significance for Asia. The region has registered remarkable economic progress in recent decades and will be the locus of the world's economic center of gravity, shifting out of the mid-Atlantic location that it occupied 40 years ago. There are great prospects on the horizon, including a weightless and digital, rather than physical, future. Equally, there are uncertainties about sustaining development, including the daunting challenges of growing inequalities, runaway climate change, and governance weaknesses. The payoffs to Asia are profound from embracing evaluative lessons on sustainability.

Policy-making is complex. Income poverty often goes hand in hand with non-income poverty. Income inequality interacts with environmental problems and climate, and these challenges themselves have roots in financial conditions. As the challenges of sustainable development take center stage in governance, this book will help inform policy and investment decisions with lessons of experience.

Dean, Lee Kuan Yew School of Public Policy   Danny Quah
Singapore

Preface

All across the world, a set of common issues affecting the welfare of people and the planet are hotly debated. How can countries pursue economic growth while at the same time achieving greater inclusion of people in the process of growth? How can progress be sustained without running down environmental capital and aggravating climate change, which already is proving to be game changing? What are the ways and means to strengthen institutions and good governance that cut across all aspects of economic performance?

With millions of people living in extreme poverty worldwide, the case for economic growth remains strong. What has changed is the recognition that growth alone is not going to deliver the goods. Rather, the quality of growth in terms of inclusion, environmental protection, and good governance is essential for sharing and sustaining the fruits of progress and in fact to ensure that growth continues at all.

The evaluation of efforts to generate and sustain economic growth remains a priority. With resource constraints everywhere, stretching development dollars is more important than ever. This book does not take its eye off the objective of growth but adds to it the bigger picture of inclusion, environmental protection, and governance. Taken together, these aggregative goals add up to development paths that are sustainable. In many ways, the globally endorsed Sustainable Development Goals (SDGs) capture the essence of sustainable development. The task of the book is to provide guidance on how best sustainable development might be evaluated. In doing so, we rely on elements from three complementary approaches or strands in the discipline of economic evaluation: impact evaluation, cost-benefit analysis, and objectives-based evaluation.

First, assessments need to ask tough questions: not just whether interventions are associated with gains to someone or another, but also if they compare favorably with alternative uses of the same funds. Assessments of how well or poorly programs have been implemented need to be complemented by impact evaluation of the addition of value over and above what might have occurred without the program in question or with an alternative approach.

Second, and in a similar vein, the assessment of benefits and costs within a framework that values investments made and the likely returns over time can be extremely valuable. Cost-benefit analysis also helps in comparing projects competing for scarce resources. If impact evaluation stresses the gains from initiatives compared to what could have been, cost-benefit analysis emphasizes the investments that need to be more than recovered for the program to be justified.

Third, there are good lessons to be gleaned from across the public and private sectors about making development programs and projects relevant, efficient, effective, and sustainable. Rigorously assessing the results against the objectives laid out in projects is necessary to promote accountability in the use of scarce resources. Making these lessons available in time for use when they still matter for new initiatives is the task of independent evaluation in partnership with operational constituents.

Risk and uncertainty pervade all these analyses, as we can see from recurring financial stresses in countries around the world. However, we know that sound macroeconomic policies and financial regulations have payoffs in staving off losses suffered by millions, and inclusive growth has a better chance of being achieved if lessons of good governance are heeded. Analogously, the mounting losses from climate-related disasters are owed to environmental destruction and runaway climate change. Climate mitigation could soften the blow going forward by capping climate extremes attributable to global warming.

In getting to these aggregative pictures of economic, social, and environmental issues, this book explores the priorities, methods, and qualifications of the analytical pieces that go into the work of evaluation. Evaluations show the connectivity among actions in related fronts on the part of countries and their development partners. The value of these evaluations also lies in generating usable outputs when they are needed. Publicizing findings, just as strengthening methods, can give evidence greater impact.

This book gives a series of illustrations of what can be learned about the overarching goals and integral parts of evaluation, all related to the sustainability of development interventions. It draws on lessons from recent evaluations, which can be a useful reference for policy-makers and practitioners. In some cases there are considerable national and regional differences that inform these lessons, and there are common lessons that cut across differences as well.

Some of the most telling lessons come when specific project evaluations are seen within the context of overarching goals or directions. The most powerful impacts are revealed when evaluation interacts with economics (as well as other social science disciplines) and illustrates inter-sectoral relations and synergies. These kinds of observations can provide insight into much-needed and value-adding work that is crucial to development efforts.

The basic message of this book is that evaluation, to be useful, needs to keep its eye on sustainable development. By the very nature of sustainable development, evaluation of the complex issues calls for analyses of interconnections among socioeconomic and environmental policies with a stress on governance. But in doing so, it pays to get the analytical methods and priorities of evaluation right, as a series of examples here illustrate.

Singapore, Singapore   Vinod Thomas
                                                    Namrata Chindarkar

Acknowledgments

We would like to thank and acknowledge the contributions and substantive input of a number of colleagues. This book project would not have been possible without the financial support from the Lee Kuan Yew School of Public Policy, National University of Singapore, and the leadership and administrative support provided by Eduardo Araral and Thomas Chan Hean Boon. We are extremely thankful to Maki Nakajima for her excellent research support from conception of the book to the final manuscript. We would like to thank the participants at the Asian Development Bank workshop including Patrick Grasso, Monika Huppi, Walter Kolkma, and Marvin Taylor-Dormond. We are grateful to Alice Martha Lee, Jyotsna Puri, Veronique N. Salze-Lozac'h, Hyun Son, Stoyan Tenev, George S. Tolley, Jiro Tominaga, and Tomoo Ueda for their detailed comments on draft chapters. We are also thankful to colleagues at the Lee Kuan Yew School of Public Policy, Andrew Francis Tan and Tan Soo Jie-Sheng, for their valuable feedback. Useful comments by Tey Sovannaroth, Stuti Rawat, and Gian Sandosh Semadeni, students at the Lee Kuan Yew School of Public Policy, are also appreciated.

Contents

1	**Evaluation, Economics, and Sustainable Development**	1
	Evaluating Sustainable Development	4
	Economics-Evaluation Interaction	9
	Evaluating Components of the SDGs	13
	Inclusive Growth	13
	Environmental Protection	15
	Institutions and Governance	16
	Conclusion	18
	Bibliography	19
2	**The Spectrum of Impact Evaluations**	25
	Why Impact Evaluation?	26
	Causal Inference and Counterfactuals	27
	Estimating the Counterfactual	28
	Establishing a Theory of Change	31
	Internal Validity and External Validity	32
	Random Assignment	32
	Sampling and Validity Issues in Randomization	33
	Which Treatment Parameters Are of Interest?	35
	Need for Baseline Information	36
	Quasi-experimental Methods	37
	Difference-in-Differences	37
	Regression Discontinuity Design (RDD)	41
	Instrumental Variables (IV)	44

    Propensity Score Matching (PSM)   48
    Choosing an Impact-Evaluation Method   52
    Challenges in Conducting Impact Evaluations   57
        Technical Challenges   57
        Organizational Challenges   57
        Political Challenges   58
    Conclusions   58
    Bibliography   59

3 The Picture from Cost-Benefit Analysis   63
    Why Use CBA?   64
    Steps in Conducting CBA   65
        Identifying Benefits   68
        Identifying Costs   70
        Valuation Techniques   71
        Avoid Double Counting   73
        Computing Net Social Benefit   74
        Externalities   75
        Project Benefits and Costs   76
        Net Present Value (NPV) of Alternative Scenarios   77
        Uncertainties and Risks   80
        Factoring Social Equity   81
        Make a Recommendation   84
    Applications of CBA   84
    Conclusion   90
    Bibliography   91

4 Objectives-Based Evaluation for Accountability and Learning   95
    Objectives-Based Evaluation   96
        OBE Criteria   97
    Project Evaluations   100
        Public Sector Projects   101
        Private Sector Projects   105
    Aggregative Evaluations   107
        Sector Evaluations   108
        Country Evaluations   112
        Evaluations by Theme   115

	Conclusion	120
	Bibliography	121
5	**Conclusion and Future Directions**	125
	Evaluation of Sustainability	126
	Direct and Indirect Impacts	127
	Well-Being and Inclusion	128
	Environmental Protection	130
	Broader Goals in Asia	130
	Local and Global Public Goods	132
	Small and Big Data	134
	Conclusion	137
	Bibliography	138

Correction: Economic Evaluation of Sustainable Development C1

Index 143

Acronyms

2SLS	Two-Stage Least-Squares
ADB	Asian Development Bank
AEA	American Evaluation Association
ATE	Average Treatment Effect
ATT	Average Treatment Effect on the Treated
CBA	Cost-Benefit Analysis
DAC	Development Assistance Committee
DID	Difference-in-Differences
EBRD	European Bank for Reconstruction and Development
ECG	Evaluation Cooperation Group
FAO	Food and Agriculture Organization
GDP	Gross Domestic Product
GGLD	Good Governance and Local Development Project
GI	Gofordev Index
GPGs	Global Public Goods
HOI	Human Opportunity Index
IDEV	Independent Development Evaluation
IE	Impact Evaluation
IED	Independent Evaluation Department
IEG	Independent Evaluation Group
IEO	Independent Evaluation Office of Global Environment Facility
IEO	Independent Evaluation Office of IMF
IEO	Independent Evaluation Office of UNDP
IEU	Independent Evaluation Unit
IFC	International Finance Corporation
IMF	International Monetary Fund
INE	National Institute of Ecology

IOE	Independent Office of Evaluation
IPCC	Intergovernmental Panel on Climate Change
ITT	Intent to Treat
IV	Instrumental Variables
JGY	Jyotigram Yojana
LATE	Local Average Treatment Effect
LGU	Local Government Units
LLN	Law of Large Numbers
MDBs	Multilateral Development Banks
MDGs	Millennium Development Goals
MTO	Moving to Opportunity
NONIE	Network of Networks for Impact Evaluation
NPV	Net Present Value
O&M	Operations and Maintenance
OBE	Objectives-Based Evaluation
ODA	Overseas Development Assistance
OECD	Organisation for Economic Co-operation and Development
OECD-DAC	Organisation for Economic Co-operation and Development-Development Assistance Committee
OED	Operations Evaluation Division
OVE	Office of Evaluation and Oversight
PM	Priority Municipality
PPAR	Project Performance Assessment Report
PPPs	Public-Private Partnerships
PSM	Propensity-Score Matching
PSM	Public Sector Management
RCTs	Randomized Controlled Trials
RDD	Regression Discontinuity Design
SDGs	Sustainable Development Goals
WHO	World Health Organization
WTP	Willingness-to-Pay
WWF	World Wildlife Fund

List of Figures

Fig. 1.1	Interactions among sustainable development issues. (Source: Authors' illustration)	5
Fig. 1.2	Development process and the results chain. (Source: Authors' illustration)	10
Fig. 2.1	Before-and-after comparison. (Source: Authors' illustration)	29
Fig. 2.2	With-and-without comparison. (Source: Authors' illustration)	30
Fig. 2.3	Random sampling and randomized assignment of treatment. (Source: Authors' illustration)	33
Fig. 2.4	DID applied to deforestation policy. (Source: Authors' illustration)	39
Fig. 2.5	Cities to the north and south of the Huai River. (Source: Chen et al. 2013)	42
Fig. 2.6	Fitted values from RDD estimation. (Source: Chen et al. 2013)	43
Fig. 2.7	Frequency distribution of treated and untreated units on common support. (Source: Capuno and Garcia 2010)	50
Fig. 2.8	Protected areas established by 2000. Protected area category: strict (green), multiple use (yellow), indigenous (pink). (Source: Nelson and Chomitz 2011)	51
Fig. 3.1	Mozambique electricity system and Ribáuè district. (Source: Mulder and Tembe 2008)	67
Fig. 3.2	Consumer surplus and producer surplus for kerosene and electricity. (Source: Authors' illustration)	74
Fig. 3.3	Marginal social cost in the presence of a positive or a negative externality	76
Image 3.1	Pollution in Mexico City. (Credit: Usfirstgov/CC BY)	87
Image 3.2	Gray water reuse facility at secondary school. (Credit: The Sustainable Sanitation Alliance Secretariat/CC BY)	89

List of Figures

Fig. 4.1	OBE of development finance. (Source: Authors' illustration)	101
Fig. 4.2	IEG ratings for outcome, sustainability, and institutional development (approval year fiscal 1995–2006). (Source: IEG 2007)	112
Fig. 4.3	Share of yearly project amounts supporting inclusive growth by pillar, 2000–2012. (Source: IED 2014b)	117
Fig. 4.4	World Bank forest activities before and after the 2002 Forest Strategy by country. (Source: IEG 2013)	119

LIST OF TABLES

Table 1.1	Development results and ADB profitability ratings	14
Table 1.2	Success rates by environmental-safeguards category	15
Table 1.3	Public sector reform lending associated with higher governance scores, 1999–2006	18
Table 2.1	Calculating the impact in DID method	39
Table 2.2	Comparison of key features of empirical evaluation methods	54
Table 3.1	Calculation of NPVs for alternative scenarios (in million US$)	79
Table 3.2	How prices shape decisions	80
Table 3.3	Illustration of NPVs for changes in market wages and time saved	81
Table 3.4	Calculation of NPVs with distributional weights (in million US$)	83
Table 3.5	Summary of annualized cost and benefits (in Indian Rupees)	90
Table 4.1	Criteria and ratings of OBE: example of the World Bank and ADB	97
Table 4.2	Assessment of community water supply and sanitation project in Sri Lanka	103
Table 4.3	IEG ratings of completed primary-education projects by year of approval	109
Table 4.4	Outcomes by enrollment objective for complete primary-education projects	109
Table 4.5	Outcomes by objective for complete primary-education projects	110
Table 4.6	IEG ratings of projects by exit year, fiscal 1992–2006 (transport sector projects versus all others)	111
Table 4.7	Sector-wise rating in country evaluation	113

Table 4.8	Assessment of the country's strategic agendas and special priorities	114
Table 4.9	Differences between the 1991 and 2002 World Bank Group Forest Strategies	118

List of Boxes

Box 1.1 Sustainable Development Goals   7
Box 3.1 CBA of Fuel Quality Improvement in Mexico (Rojas-Bacho et al.
         2013)   86
Box 3.2 CBA of Water Projects in India (Labhasetwar 2013)   88

CHAPTER 1

Evaluation, Economics, and Sustainable Development

Abstract This introductory chapter opens the discussion on applying economic tools for evaluating sustainable development. It sets out the focal argument of this book that the current approach to development evaluation, which primarily assesses the value added of growth with only a cursory glance given to issues of inclusion, environmental stewardship, and good governance, needs to evolve to more thoroughly account for these critical issues.

Keywords Economic growth • Inclusion • Environment • Governance • Climate change

> *One of the great mistakes is to judge policies and programs by their intentions rather than their results.*
> Milton Friedman

Evaluation has a rich history in informing work on economic development. Both countries and financiers have used evaluations to improve their work on development projects, individual sectors, and sometimes the

---

The original version of this chapter was revised. An erratum to this chapter can be found at https://doi.org/10.1007/978-981-13-6389-4_6

© The Author(s) 2019
V. Thomas, N. Chindarkar, *Economic Evaluation of Sustainable Development*, https://doi.org/10.1007/978-981-13-6389-4_1

economy. Multilateral, bilateral, and United Nations development finance agencies have funded evaluations of their work in countries and regions over the years, and most now have evaluation offices, many of them independent in their mandate (Picciotto 2013). Colombia, Mexico, South Africa, the United Kingdom, and the United States are among the growing number of nations that have strengthened evaluation capacity over the years (Thomas and Luo 2012).

A premise in all this is that different methods of evaluation in varying contexts can help make development programs more effective (OECD 2010). Evaluation assumes great importance when competition for scarce resources increases. In times of crises, such as the 2008 global financial crisis, there was demand for information on how government-funded programs were performing. Policy-makers and the public need to know which programs are likely to achieve a high development impact and which are not, and evaluation can try to provide that, as well as lessons for improving programs' performance.

Different approaches have tried to track the results of interventions. Impact evaluation (IE) has been increasingly applied to programs in social areas, such as education, health, and social protection (Sabet and Brown 2018). Infrastructure investments, for example, in energy, transport, and water supply, have usually been put to the test using cost-benefit analysis (CBA). Development agencies have assessed how well programs are delivering on the objectives they set out using objectives-based evaluation (OBE).

Evaluation intersects with economic analysis when assessing economic development and also related social policy objectives such as investing in people. Economic analysis has long been applied to influence development policies at the macroeconomic and microeconomic levels (IEG 2010, 2012). There is a vast body of evidence on the economic costs and returns of having greater openness in trade policy (see, e.g., Lukauskas et al. 2013). Analysis of the effects of trade liberalization on agriculture, industry, or services provides grounds for policy reform. Similarly, a great deal of empirical work has tried to shed light on the economic returns to individuals or households from having more education (for instance, Hanushek et al. 2006; World Bank 2006).

One overarching objective of this book is to illustrate how evaluation and economic tools can be applied more meaningfully with reference to how development goals are being furthered. Development is tied to human, social, and environmental concerns and impacts on future generations. This idea of "sustainable development" encapsulates the principal considerations of policy and action. By the very nature of sustainable development, environmental protection and climate change become key aspects of economic growth.

Development evaluation must find ways to assess sustainable development and not just aspects of economic growth, as is often done. Specifically, the socio-political and economic landscapes of nations need to factor in goals of promoting greater social inclusion, environmental protection, or better governance. By bringing economics and evaluation together, we can better see the results of interventions through the lens of sustainable development.

A critical question that arises is the "evaluability" of complex sustainable development issues such as climate change, social inclusion, and good governance. There is recognition that these issues present major challenges to traditional objectives-based evaluative enquiry (McGrail 2014). Evaluating projects and interventions aimed at addressing them thus requires reformulating evaluation goals and objectives, rethinking the framing and design of evaluations, and blurring of evaluation boundaries from being intervention-focused to being more aggregative.

Explicating the connection between evaluation and economic policy analysis is another overall objective of this book. The payoffs to forging connections between economics and evaluation can be high, but opportunities have not been adequately seized, and economics and evaluation have not been brought together sufficiently. Often bureaucratic motivations have kept work in separate disciplines. Limitations of methods of analysis and data availability have also stood in the way of stronger interactions between the disciplines.

The interlinkages among strands of policy issues are complex, both with respect to the challenges they pose and the opportunities they present. This is not only the case for broader issues in economic development but also for interventions, which can be individual projects or national-level programs or policies. Economic motivations interact with social and political forces in development interventions. For instance, trade liberalization connects at the same time with sources of welfare gains and welfare losses to specific groups in varying degrees, which then affect the feedback on trade liberalization policies and democratic decisions made.

Bringing economics and evaluation together can create better interactive links among policy areas, which, when exploited, illuminate aggregative issues of concern in development. Individual project or sectoral analyses are valuable, and they are essential building blocks for assessing crosscutting areas. However, if they are relied on to the exclusion of other tools, as is often done, evaluation can fall short of its promise of informing policy directions.

Evaluating Sustainable Development

As development challenges become more complex, replicating what worked in the past—even projects rated as highly successful—is no guarantee of continuing success. Continuing development problems, such as groundwater depletion, quality of education, and income distribution, remain intractable. Previous solutions may no longer suffice given a changing physical, social, and economic environment. New problems, such as rising incidence of non-communicable diseases, environmental degradation, and climate change, add to the premium for innovation in projects and development portfolio.

A recent review summarizes eight challenges in development that evaluation would do well to confront and address (Basu et al. 2016). These issues resonate as priorities for development evaluation: economic growth as a means to well-being; inclusive growth; environmental care, including climate change; market-state balance and regulation; macroeconomic stability; technological change; social norms; and changes in global economic forces.

Much of evaluation of projects and programs directly or indirectly deals with the impact on efficiency and effectiveness of economic growth. There is a great deal of good project evaluations as well as some evaluation studies considering the economy-wide impact of a financial crisis (e.g., IEO 2014), trade liberalization, or other changes on growth. But the current approaches primarily deal with value addition in terms of economic growth, with only a limited look at income distribution, environmental protection, and good governance.

The challenge for evaluation is integrating environmental, social, and institutional aims while assessing growth. The United Nations Sustainable Development Goals (SDGs) capture these dimensions in the form of targets (United Nations 2016). Following SDGs and empirical results on development attributes (see, e.g., World Bank 1991, United Nations 2016), this book takes sustainable development to mean a combination of economic growth, social inclusion, and environmental stewardship underpinned by good governance.

As illustrated in Fig. 1.1, these developmental issues are distinct but also interact with each other.

By the nature of the measurement of improvements, economic growth is rightly captured in evaluation. Broadening of the focus of evaluation from the pace of growth to the quality and impact of growth calls for

Fig. 1.1 Interactions among sustainable development issues. (Source: Authors' illustration)

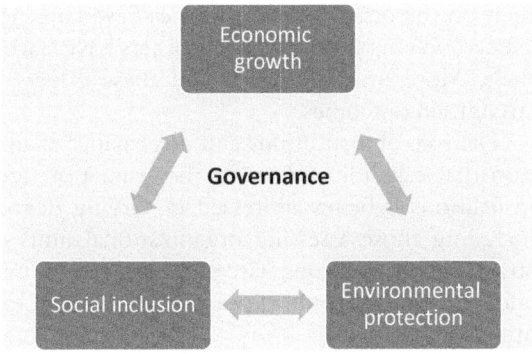

policy to inter alia address inclusive growth. Policy-making would want to see if people are left behind, unemployment is tackled, and inadequate health care and lack of access to education are dealt with. These topics merit evaluation, as evaluation agendas of various organizations indicate.

Measurement and analysis of extreme poverty have received considerable attention, in part because poverty reduction mirrors income growth (holding income distribution constant). The period of the Millennium Development Goals (MDGs) saw a sizable decline in extreme poverty (Ravallion 2013), but serious challenges remain especially as hazards of nature or food price shocks can easily put vulnerable populations back into poverty. Income distribution has worsened in many countries although evaluations do not take up distributional issues often.

Environmental protection is central to sustainable development. At the national level, income growth that comes at the cost of environmental damage, such as air and water pollution, is proving to be unsustainable. Beijing and New Delhi, the capital cities of the world's most populous countries, suffer from dangerously high levels of air pollution. Globally, climate change is a threat to health, livelihoods, and habitats. Climate change mitigation and adaptation need to be integral to development policy, and their evaluation, difficult as it may be, needs to become part of the tool kit.

The interaction among attributes of sustainable development comes through strongly in the case of climate change. For instance, for infrastructure investments to generate lasting growth, they need to take into account climate effects. Energy policies impact the adoption of renewable and clean energy supply on the one side and the use of polluting fossil fuels on the other. A few countries have launched cap-and-trade schemes for carbon emissions, and a few others have started levying a carbon tax on fuels. More must be done, and these efforts need to be evaluated to strengthen outcomes.

The role of institutions and governance is another overarching dimension that calls for updates to the evaluation agenda. The functioning of institutions is being addressed in varying degrees by evaluation groups, including those assessing organizational units or corporate entities and strategies or directions. Greater rigor can be introduced into such work, and the role of public goods and market externalities given greater attention.

A continuing question is if the pursuit of sustainable development will come at the expense of economic growth. The neglect of environmental externalities in policy-making would seem to suggest that it views economic growth on the one side and environmental care and climate action on the other as conflictive. Evaluations might shed light on the possibility that climate mitigation and other sustainability measures might contribute to continuing economic growth and, by the same token, turn them into market opportunities.

In this book, we provide evidence from the experience of International Finance Corporation (IFC) and Asian Development Bank (ADB) that economic returns and sustainable development likely go together (Chap. 4). The key point is not only that economic growth, environmental protection, and social sustainability can go hand in hand but also that it may be hard to sustain growth without social inclusion and environmental care.

Evaluating sustainable development can have important payoffs in two ways. It can help keep the eye on broader development goals, such as the SDGs. It can also help to adopt efficient and effective policies and investments to further sustainable outcomes. There is likely underinvestment in evaluating development effectiveness in general (Ravallion 2009). On top of that, the value of evaluating sustainability suggests the need for even stronger evaluation efforts.

Considering sustainability in evaluation has usually meant different things to different people. One way of thinking is whether a project or an intervention itself is sustained into the future, especially after the funding for it has ended. This of course has broader implications in that if the project is not sustained, its benefits too may not last.

A second approach is to look at this latter aspect directly, that is, assessing the extent to which the benefits of a project, program, or policy are maintained after formal support has ended. Development Assistance Committee's (DAC) evaluation framework includes sustainability as one of its five criteria of evaluation (see Chap. 4). Under this criterion, financial and institutional and sometimes environmental care too are considered.

A third approach, which is the focus of this book, is to think of the impact of a project or other forms of intervention on sustainable development (see IEU 2018 for examples). In so doing sustainable development might be taken to mean "Development that meets the needs of the present without compromising the ability of future generations to meet their own needs" (Brundtland 1987).

This approach might in part be synonymous with environmental stewardship, as this is a highly vulnerable aspect of efforts that target economic growth. But, under the SDGs, sustainability goes much further than the environment, although the stress on the environment and climate is much stronger than under the MDGs. In addition to the environment, the SDGs emphasize social inclusion and governance (Box 1.1).

The term "sustainability," which has roots in forest management, might refer to human-ecosystem balance, while "sustainable development" refers to underlying temporal processes (Shaker 2015). It would be fair to say that policies aimed at sustainable development would promote the best use of resources to help meet human needs while protecting the integrity of the natural system, which in turn is essential for future human needs to be met.

Box 1.1 Sustainable Development Goals
The Sustainable Development Goals (SDGs) of the 2030 Agenda for Sustainable Development were adopted in September 2015 at the United Nations Sustainable Development Summit and officially came into force in January 2016. The goals were based on the lessons from the Millennium Development Goals (MDGs). In effect during 1990–2015, the MDGs had established a common platform to tackle extreme poverty and hunger, universal education, and better health. During that period, extreme poverty rate dropped from 47 percent to 14 percent, the number of out-of-school children of primary school age declined from 100 million to 57 million, and child mortality dropped from 12.7 million to 6 million (United Nations 2015).

The SDGs place greater values than MDGs on building a sustainable world with environmental protection, social inclusion, and economic development. One of the new goals is to combat climate change and its impacts on public health, food and water security, migration, peace, and security. While MDGs were intended for low-income countries, the new goals cannot be achieved without the efforts of all countries including high-income ones.

The 17 SDGs are no poverty; zero hunger; good health and well-being; quality education; gender equality; clean water and sanitation; affordable and clean energy; decent work and economic growth; industry, innovation, and infrastructure; reduced inequalities; sustainable cities and communities; responsible consumption and production; climate action; life below water; life on land; peace and justice strong institutions; and partnerships to achieve the goals.

At the global level, the achievement of SDGs is monitored using the global indicator framework developed by the Inter-agency and Expert Group on SDG Indicators, prepared in March 2015 at the session of the United Nations Statistical Commission. The High-level Political Forum, established in 2012, meets annually and serves as the main platform for follow-up and review of SDGs. The Forum offers a means to monitor the progress in each country and region, exchange the best practices, and to foster international cooperation.

A recent report (United Nations 2018) states that, while there has been some progress in the three years after the SDGs were implemented, progress has not been rapid enough for the targets to be achieved by 2030. It reiterates that the challenges in achieving the ambitious goals are interrelated and integrated approaches need to be adopted. For example, proper management of wastewater is closely related with public health and the environment. It also highlights the crucial gaps in data from national statistical and data systems to monitor the progress toward the goals (United Nations 2018). There have also been some efforts to assess the synergies and trade-offs among the SDGs (Pradhan et al. 2017).

Economics-Evaluation Interaction

Economic thinking in evaluation design pertains to reflecting more analytically about the relationship between the objectives of a program or intervention and the results. How a project or intervention is expected to achieve results depends on the underlying assumptions—on the validity of what economic theory and practice expect. An evaluation based on economic thinking begins by laying out a chain linking inputs to outputs, outcomes, and impacts for a given project.

To answer why the intervention worked or did not work, mapping out the results chain to test the underlying assumptions is key. Many of the events or conditions that are assumed to produce the desired outcomes might not be in place. Nor might the interventions function as expected, particularly in view of the growing complexity and interrelatedness of development programs. Assumptions need to be identified and tested in relation to the macroeconomic and political environments. Evaluation can unbundle the theory of change to review how an intervention might convert inputs and outputs into outcomes and impacts. Theory of change is an approach for evaluation grounded in the mechanics of social change, looking at goals and mapping backward to identify preconditions.

To be effective, evaluation needs to consider the links connecting inputs to outputs—and to outcomes and impacts (Fig. 1.2). This requires focusing on identifying what might be the right results, getting the appropriate measures, and providing lessons to enhance development effectiveness. To ensure some degree of objectivity, the results might revolve around commonly accepted and well-articulated development goals, such as the SDGs. The development community has tried to move from a focus on inputs and outputs to a consideration of outcomes and impacts, as has been seen in a series of events including the 2002 International Conference on Financing for Development in Monterrey, Mexico, which established the MDGs, and the 2008 Forum on Aid Effectiveness in Accra, Ghana. The adoption of the SDGs underscored the focus on getting results on the ground.

The focus on outcomes and impacts draws attention to the vital links in the results chain and to the complexity of attributing outcomes to particular inputs. Many factors influence results, including conditions outside the domain of the interventions. The findings of evaluations refer to and intersect with the full process of the development, from inputs to outputs, to outcomes, and to impacts relating to the interventions. By considering the development process in designing an evaluation, findings can have value not only retrospectively but also in real time and prospectively.

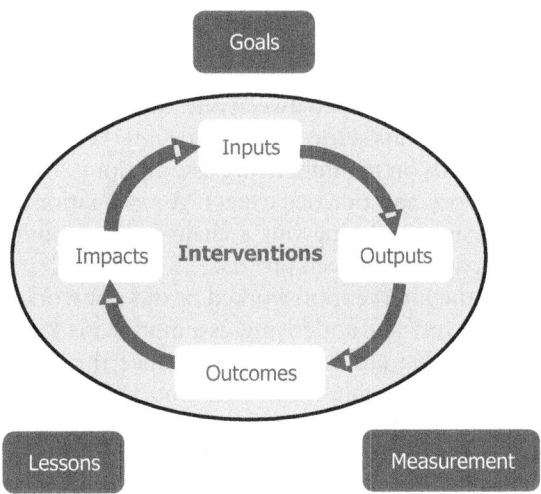

Fig. 1.2 Development process and the results chain. (Source: Authors' illustration)

Some evaluations have faced the criticism that they do not deal with unintended effects and complexities. Not only direct effects, such as the contribution of investment to economic growth, but also indirect effects, such as the influence of improved access to water and sanitation on girls' education, are to be considered. These latter links are often not considered, let alone quantified. The results chain must therefore consider the intended consequences of development activities and also the unintended impacts, such as the social dislocations caused by a road project or a water project, which can be just as important in urban and rural settings (Tolley et al. 1979). It is not enough to measure only the intended results, because the unintended ones may provide unexpected benefits or costs. Unintended results can provide rich sources of learning for future activities and check on current ones.

Evaluations can bring out complementary factors and synergies for development success. For instance, links between the public and private sectors through public-private partnerships (PPPs) could offer new approaches to service delivery and prove to be key to outcomes. Take, for example, PPPs in the agricultural sector. Given the private nature of agricultural activities and the public-good nature of agricultural services, particularly agricultural research and extension, the extent to which interventions link government and private producers makes a difference for performance of the agricultural sector. Transport projects in various settings are seen to improve inclusion if they are linked with programs addressing education and health care. Observations of rural road projects in Bangladesh point to gains when investment takes place in related areas such as education, health, and financial literacy. Yet another example is education policy. Education investment pays off when coupled with labor-market reforms to support job creation, especially for the lower-income strata.

Measurement is another important aspect of evaluation. Independence, objectivity, and the impartiality of data are themselves a big part of the validity and value of evaluations. By setting clearly measurable objectives, analysis can focus on achievements that can be independently verified.

Often, evaluation is thought to be constrained by the lack of adequate data and information. But it is part of the evaluative process to seek and ensure sufficient data that are credible to lead to the evaluability of projects, programs, and interventions. The appearance of big data can be a potential aid to this endeavor, as Chap. 5 suggests.

In development economics, various empirical methods, including econometric analysis, measure the effectiveness of interventions. Evaluation has evolved with some dominant approaches and several strands of analytical methods tailored to specific situations, including qualitative assessments (AEA 2004; ECG 2010; IED 2014).

IE, as elaborated in Chap. 2, measures the change attributable to a program or intervention and tries to answer the question—what difference does the program make? It considers the counterfactual, which could be pre- versus post-program situation, or with and without the intervention. This approach can help assess the effects of programs that seek to ensure greater social inclusion and environmental protection (Croke et al. 2017). The much-cited example is the case of social protection programs, where an evaluation delineates the impacts of conditional cash transfers.

CBA, detailed in Chap. 3, is a long-standing economic tool of analysis especially for infrastructure projects, but it can be put to better and wider use to assess social and environmental projects as well. It does well when data on costs and benefits of the intervention can be gathered, which is usually easier for the so-called hard sectors like infrastructure (see, e.g., Harberger 1976; Boardman et al. 2006). The evaluated projects are either in the public sector or in the private sector. Its framework provides for valuation techniques to account for externalities such as pollution and congestion as well.

Multilateral development banks heavily use assessments of accomplishments against agreed-on or planned goals, in what is often referred to as objectives-based evaluation, as discussed in Chap. 4. The approach is applied for public sector as well as private sector projects. There are well-established criteria to judge project success: relevance, efficiency, effectiveness, development impact, and sustainability (DAC 1991, 2018). Often, ratings on each of these criteria are aggregated to assess overall performance.

The bulk of evaluations have a project and micro, and at most a sectoral and thematic, focus. Cross-linkages and macro-aggregations are not often done, even when actions taken at higher policy levels are decisive factors in individual project-level success. For example, while individual analysis of environmental projects is valuable, the government's environmental regulation might have overriding importance.

If evaluation is to contribute to improving sustainable development outcomes, then it must straddle project and sectoral boundaries and make calls at the macro or aggregative levels. It must assess impacts on aggregative goals such as inclusive growth, environmental care, and good governance. Doing so requires evaluation to work closely with the economics discipline. There are risks in doing so, but the rewards would be high.

Rather than thinking of these tools and disciplines as alternatives, they can be considered as part of a rigorous framework that mixes methods depending on the issues at hand. Crucial to this approach would be the identification of high-priority objectives and issues. IE can be applied more widely than at present, not only to social policies but also to urban development, infrastructure, and climate change policies. We must also take CBA more seriously and not let data limitations discourage its use. OBE would benefit from deepening linkages with economic analysis and incorporating evidence from complementary approaches.

Evaluations and available data often lead to findings that confirm what is known. Here, its value lies mainly in summarizing lessons and, perhaps, in suggesting improvements for future interventions. But some evaluations generate unexpected results that question the assumed connections between actions and desired outcomes, including the critical assumptions and context for the underlying theory of change implicit in the activity.

By pointing out crucial but neglected areas and providing timely information to change development thinking and guide policy decisions, evaluations can push policy interventions from a generally accepted but perhaps ineffective state of inertia to a more beneficial course. There is a premium in evaluation taking on cutting-edge issues in sustainable development, even when data and conceptual aspects constrain the analysis.

Evaluating Components of the SDGs

In the context of sustainable development, Thomas et al. (2000) discussed the need to consider the quality of growth, in addition to its quantity, in terms of social inclusion, environmental stewardship, and the accompanying governance. Sachs (2012) signaled the importance, in addition to economic growth, of inclusion, the environment, and good governance in thinking about sustainable development. The goals and targets under the SDGs can be laid out within these overarching aims. Progress along these axes can be tracked, monitored, and assessed (Kharas et al. 2018).

But these issues present challenges to evaluation. In particular, evaluation priorities and methods have not kept pace with the needs of assessing outcomes in sustainable development. We need to step up evaluation efforts at several levels, improving frameworks; methods of analysis; and relevant and practical applications, conclusions, and recommendations. We now take up some illustrations of how evaluation might view the principal components of the SDGs.

Inclusive Growth

There is a growing recognition within countries that growth that is inclusive is vital for how it impacts people's well-being and for continuing economic growth itself. Empirical studies suggest that not only does higher inequality tend to limit the impact of growth on absolute poverty but also that countries with high inequality may experience rising poverty despite good growth prospects (Ravallion 1997). Piketty (2014) shows that as economies develop, the uneven distribution of skills and education of the workforce promotes inequality, and inequality continues to increase unless some measures are taken (see also Lakner and Milanovic 2013).

Redistribution policies that use taxes and transfers, while politically sensitive, are still the predominant tools used to address inequality. One way to assess whether redistribution policies promote inclusive growth or not is to distinguish market inequality (before taxes and transfers) from net inequality (after taxes and transfers), as done by Ostry, Berg, and Tsangarides (2014). Their empirical analysis suggests that the impact of redistribution on inclusive growth is generally positive except for countries where difference between the market and net inequality is extremely large (see also Sabot et al. 2016).

The desire to increase economic growth remains the principal driver of policy, but there are good grounds for building inclusion into the design and implementation of projects intended to help raise economic growth. In the past, social inclusion and environmental protection were viewed as good to have but their pursuit presented unacceptable trade-offs to economic growth. However, some results have shown that projects with objectives incorporating inclusive growth have performed well compared to those that do not (IED 2016).

A review of 94 private sector projects at ADB since 2006 seems to suggest that development results and investment profitability are not necessarily incompatible. Table 1.1 shows the association (not causality) between a project's profitability and its development results. It suggests that 56 of the 94 projects evaluated (60 percent) were rated by the criteria used as both profitable and successful in contributing to development. Earlier exercises done at the World Bank on larger samples showed a similar association (see Chap. 4).

In these private sector interventions, projects have tried to address inclusive growth through two main channels. The majority of them have invested in areas where there is a constraint on inclusive growth. These investments benefit people at the bottom of the pyramid, providing infrastructure and financial services. In addition, there are inclusive business transactions that work with businesses that provide services to the poor, primarily employing people from disadvantaged groups or including the poor in their supply chains.

The recognition of the importance of inclusive growth raises several challenges, an important one being the management of actual or perceived trade-offs. One example is evaluating the cost involved in expanding the reach of education and health services as well as social protection, all of which will aid inclusion. CBA is particularly suited to weigh the additional

Table 1.1 Development results and ADB profitability ratings

ADB profitability		Development results	
	High	9	56
		10%	60%
	Low	23	6
		24%	6%
		Low	*High*

Source: IED (2016)

expenditures against the stream of benefits accruing from broader participation of people in the growth process.

Environmental Protection

Environmental protection remains a contentious, complex, and dynamic area with a number of perceived or actual trade-offs to be considered. One example is food security: there is the need to increase areas under cultivation while at the same time ensuring sustainable forest use and conservation. Another example is the pressure to develop fossil-fuel energy to power growth which conflicts with controlling pollution and minimizing damages to human health and mitigating climate change.

However, there is growing evidence that sustained growth will not be possible in the future without tackling environmental degradation and climate change. Climate change is the greatest known threat to sustainable development, and its impacts go far beyond natural disasters (Stern 2006). The costs of climate-related disasters in many disaster-prone countries like Bangladesh, Cuba, Haiti, Thailand, and the Philippines are staggering, and they weigh down on economic growth.

A concern for environmental protection is sometimes seen as an impediment for delivering efficient and effective projects as well as for supporting rapid growth. But evaluation results do not seem to endorse this concern. A review of the success rates of projects with environmental-safeguards categories shows an interesting association (not causality). Those which needed more substantial environmental safeguards (because of higher risks) performed better in terms of estimated project success rates (Table 1.2). Projects are labeled Category A when they are likely to have significant environmental risks.

These projects require an environmental-impact assessment and an environmental-management plan, as well as an elaborate process of consultation and coordination, bringing higher levels of complexity and risk.

Table 1.2 Success rates by environmental-safeguards category

	Category A	Category B	Category C
Success rate	84%	62%	57%
Number of projects rated	64	263	166

Source: IED (2016)

Despite these added challenges, the success rate of these projects is higher than for projects with more limited (Category B) and minimal or no potential environmental impacts (Category C). This may suggest that the extra scrutiny Class A projects get on environmental grounds may have positive spillover effects on the broader design and implementation.

Evaluation of climate change needs to account for more than goods and services that can be monetized. Complementing CBA with additional decision-making tools can make evaluations more comprehensive and provide more robust insights. Many of these tools, such as green accounting methods, are in principle available for better valuations of natural capital. The availability of data is usually a constraint to such valuations. But they are important considering that when the destruction of natural capital is not accounted for, growth prospects are likely inflated.

IE has been applied to assess policies for mitigating climate change and environmental degradation. Examples include an evaluation of Brazil's deforestation control policies, suggesting that when a municipality is designated as a priority, deforestation rates within about 50 kilometers of its boundaries decrease from improved monitoring, but rates farther away increase from displacement of illicit activity (Slough and Urpelainen 2018). Another evaluation of community-based forest management in Ethiopia (Takahashi and Todo 2012) found that the forest area managed by forest associations declined more in the year of establishment than forest areas with no association, perhaps from "last-minute" logging. But on average, the forest area of the forest associations increased by 1.5 percent in the first 2 years, whereas those not managed as part of an association declined by 3.3 percent.

Evaluators have been slow in applying economic evaluation tools to environmental issues, but it is now urgent that the discipline comes to grips with it. It is only with a swift policy response based on sound evidence that countries can highlight and address issues threatening environmental protection and achieve sustainable development.

Institutions and Governance

There is no universal strategy for pursuing a triple bottom line of growth, inclusion, and environmental protection, but having better institutions and good governance, which cut across all these areas, helps. That puts the evaluations of institutions, corporate structures, incentives, and performance at the center. Global measures of good governance vary a great deal across regions and countries and over time (Kaufmann et al. 2009). For example, Southeast Asia fares poorly in its control of corruption, while in East Asia, the gaps are wide for voice and accountability—an indicator which captures perceptions of the extent to which citizens can participate in policy-making processes and the accountability of governments. South Asia, meanwhile, ranks low in political stability.

Good governance could lead to sustainable development through various channels. It plays a critical role in promoting inclusive growth by ensuring that public services actually reach the poor and disadvantaged. Development practitioners know the deleterious effects on health and education of absenteeism of doctors and teachers, especially in remote rural areas. Consider also the crisis of climate change: the elimination of fossil-fuel subsidies has long been advocated as a means to cutting back high-carbon energy and freeing up funding for green-energy projects. However, their implementation comes up against the political economy of such reform.

But there is also evidence that even modest improvements pay off. One review (IEG 2011) indicates that the achievement of country outcomes was correlated with country governance, measured by the Country Policy and Institutional Assessment (CPIA) data. Just four out of nineteen programs in countries with low CPIA governance scores (3.2 or less) had satisfactory outcomes, compared with 75 percent in those with high CPIA governance scores. Generally speaking, when governance is off course, projects seem to do poorly.

Governance projects supported by external financial agencies usually fall into the categories of public sector management (the largest segment), financial management, civil-services reform, and anti-corruption activities. An OBE of the success rates of these projects shows that they generally fall below the overall average performance, signaling the difficulty of working in the governance area (IEG 2008).

Some evidence points to the potential for governance projects to work. IEG 2008 showed a large difference in estimated governance scores between countries that borrowed from the World Bank for public sector reform and those that did not (Table 1.3). Overall, borrowers had a 73 percent improvement rate in terms of countries that improved the CPIA and non-borrowers a 48 percent improvement rate in this estimated governance score. Across regions the correlation of public sector reform lending with changes in governance scores varied sizably. Europe and Central Asia had the highest rate of improvement for countries getting such lend-

Table 1.3 Public sector reform lending associated with higher governance scores, 1999–2006

Region	With World Bank public sector reform lending		Without World Bank public sector reform lending	
	Percent	Number	Percent	Number
Sub-Saharan Africa	70	30	47	15
East Asia and the Pacific	70	10	56	9
Europe and Central Asia	90	20	86	7
Latin America and the Caribbean	75	20	25	8
Middle East and North Africa	57	7	0	2
South Asia	50	6	0	1
Total	73	93	48	42

Source: IEG (2008)
Note: Entries show the percent and number of countries with an improvement in the average CPIA during 1999–2006

ing (90 percent), and the rate of improvement for non-borrowers is almost as high. The explanation might lie elsewhere, for instance, European Union accession.

Service delivery is a key aspect of good governance. Developing mechanisms and harnessing information technology to improve information sharing, transparency, and civic participation have the potential to improve the delivery of services. Recent IEs have attempted to unravel the effect of public service delivery on achievement of the SDGs. Kingdon and Muzammil (2013) find that unionization makes public school teachers less accountable toward student performance and lowers incentives to put in effort on student learning, thus resulting in low test scores. Yamada, Sawada, and Luo (2013) find that improved health service delivery owing to timely payment of wages is negatively associated with absenteeism among public health workers in Lao PDR. Such findings can provide a picture of what needs to be done to improve public service delivery.

Conclusion

A great deal of progress has been made in applying evaluation tools to the assessment of individual projects, programs, or interventions. Projects and programs remain the building blocks for achieving broad development goals. But there is a gap in linking the economics of these actions with the evaluation findings and in connecting the dots to see how overall development goals are being achieved.

This book encourages more integration of economics and evaluation analytics. Historically, the evaluation field grew out of the need to assess social programs and support legislation of the programs, as, for example, in the 1960s in the United States. There has been a focus on psychometrics, surveys, and data, but not the economics of the issues being tackled. The major encounter between economics and evaluation has been in development evaluation, and it has not been an easy one. IE opens the door for a much greater economics-evaluation integration. CBA too has the potential for expanding such connectivity. OBE can provide the platform to integrate methods of economic evaluation with other evaluation techniques for a more comprehensive assessment.

The book also pushes evaluations to go from being value-neutral to embracing a more policy-oriented, and at the same time rigorous, role. To make this transition, evaluations can be done against well-articulated goals, such as the SDGs. It would be valuable to introduce the issues of inclusive growth and environmental protection underpinned by good governance as the three overarching goals encompassing the SDGs. If evaluations were to adopt the SDGs as a measuring rod, it would be possible to get good comparative analysis of what is working in development. Improvements in evaluation techniques are also essential. In particular, evaluation techniques need to be adaptive, sensitive to complexity, and amenable to feedback and replication.

An important goal for achieving sustainable development is capacity building. In the context of evaluation, it refers to developing evaluation capacity not only among established institutions but also among new enterprises on a country-competency level. Contributing toward this goal is the larger objective of this book.

Bibliography

AEA (American Evaluation Association). 2004. Guiding Principles for Evaluators. Fairhaven, MA: AEA.

Bamberger, Michael, Jos Vaessen, and Estelle Raimond. 2016. *Dealing with Complexity in Development Evaluation: A Practical Approach*. California: Sage.

Basu, Kaushik, Francois Bourguignon, Justin Yifu Lin, and Joseph E. Stiglitz. 2016. "A New Year's Development Resolution." https://www.project-syndicate.org/commentary/update-development-policy-inequality-by-kaushik-basu-et-al-2016-12?barrier=accessreg.

Boardman, Anthony E., David H. Greenberg, Aidan R. Vining, and David L. Weimer. 2006. Cost-Benefit Analysis: Concepts and Practice, Vol. 3. Upper Saddle River, NJ: Prentice Hall.

Brundtland, Gro Harlem. 1987. Our Common Future: Report of the 1987 World Commission on Environment and Development. Oslo: United Nations.

Cameron, Drew, and Jorge Miranda. 2015. "Trends in Impact Evaluation: Did We Ever Learn?". http://blogs.3ieimpact.org/trends-in-impact-evaluation-did-we-ever-learn/.

Castelló-Climent, Amparo. 2010. "Inequality and Growth in Advanced Economies: An Empirical Investigation." *Journal of Economic Inequality* 8 (3):293–321.

Cotterman, Ron. 2016. "The Sustainability Paradox: Why Business Leaders Need to Evolve Their Approach." https://sealedair.com/insights/sustainability-paradox?gclid=EAIaIQobChMI6Iqek4K43AIVgTqBCh2bSgiHEAAYAiAAEgIYcfD_BwE.

Croke, Kevin, Eric Hsu, and Michael Kremer. 2017. "More Evidence on the Effects of Deworming: What Lessons Can We Learn?." *American Journal of Tropical Medicine and Hygiene* 96 (6):1265–1266.

Dabla-Norris, Era, Kalpana Kochhar, Nujin Suphaphiphat, Frantisek Ricka, and Evridiki Tsounta. 2015. *Causes and Consequences of Income Inequality: A Global Perspective, IMF Staff Discussion Note*. Washington, DC: International Monetary Fund.

DAC (Development Assistance Committee). 1991. The DAC Principles for the Evaluation of Development Assistance. Paris: OECD.

DAC (Development Assistance Committee). 2018. "DAC Criteria for Evaluating Development Assistance." OECD. http://www.oecd.org/dac/evaluation/daccriteriaforevaluatingdevelopmentassistance.htm.

ECG (Evaluation Cooperation Group). 2010. Good Practice Standards on Independence of International Financial Institutions' Central Evaluation Departments. Manila: Evaluation Cooperation Group.

Halter, Daniel, Manuel Oechslin, and Josef Zweimüller. 2014. "Inequality and Growth: The Neglected Time Dimension." *Journal of Economic Growth* 19 (1):81–104.

Hanushek, Eric A., Stephen Machin, and Ludger Woessmann, eds. 2006. *Handbook of the Economics of Education*. Amsterdam: North Holland.

Harberger, Arnold C. 1976. *Project Evaluation*: University of Chicago Press.

Heider, Caroline. 2017. "Rethinking Evaluation—Sustaining a Focus on Sustainability." IEG. http://ieg.worldbankgroup.org/blog/rethinking-evaluation-sustaining-focus-sustainability.

IED (Independent Evaluation Department). 2014. Evaluation for Better Results. Manila: ADB.

IED (Independent Evaluation Department). 2016. Annual Evaluation Report. Manila: ADB.

IEG (Independent Evaluation Group). 2008. Public Sector Reform: What Works and Why—An IEG Evaluation of World Bank Support. Washington DC: World Bank.
IEG (Independent Evaluation Group). 2009. Knowledge for Private Sector Development. Washington DC: World Bank.
IEG (Independent Evaluation Group). 2010. The World Bank Group's Response to Global Economic Crisis (Phase I). Washington DC: World Bank.
IEG (Independent Evaluation Group). 2011. Results and Performance of the World Bank Group: IEG Annual Report 2011. Washington DC: World Bank.
IEG (Independent Evaluation Group). 2012. World Bank Group Response to the Global Economic Crisis. Washington DC: World Bank.
IEO (Independent Evaluation Office). 2014. IMF Response to the Financial and Economic Crisis. Washington DC: International Monetary Fund.
IEU (Independent Evaluation Unit). 2018. Independent Review of the GCF's Results Management Framework, Evaluation Report No. 1/2018. Songdo, South Korea: Green Climate Fund.
IPCC (Intergovernmental Panel on Climate Change). 2014. Climate Change 2014: Impacts, Adaptation, and Vulnerability. Geneva, Switzerland: IPCC.
Kanbur, Ravi. 2016. "Economic Growth and Poverty—the Inequality Connection." https://en.unesco.org/news/ravi-kanbur-economic-growth-and-poverty-inequality-connection.
Kaufmann, Daniel, Aart Kraay, and Massimo Mastruzzi. 2009. Governance Matters VIII: Aggregate and Individual Governance Indicators 1996–2008. *Policy Research Working Paper no. WPS 4978.* Washington DC: World Bank.
Kharas, Homi, John McArthur, and Krista Rasmussn. 2018. "Counting Who Gets Left Behind: Current Trends and Gaps on the Sustainable Development Goals."
Kingdon, Geeta, and Mohd Muzammil. 2013. "The School Governance Environment in Uttar Pradesh, India: Implications for Teacher Accountability and Effort." *Journal of Development Studies* 49 (2):251–269.
Lakner, Christoph, and Branko Milanovic. 2013. Global Income Distribution: From the Fall of the Berlin Wall to the Great Recession. *Policy Research Working Paper no. 6719.* Washington DC: World Bank.
Lukauskas, Arvid, Robert M. Stern, and Giani Zanini, eds. 2013. *Handbook of Trade Policy for Development.* Oxford: Oxford University Press.
McGrail, Stephan. 2014. "Rethinking the Roles of Evaluation in Learning how to Solve 'Wicked' Problems: The Case of Anticipatory Techniques used to Support Climate Change Mitigation and Adaptation." *Evaluation Journal of Australasia* 14 (2):4–16.
Morgan, Trevor. 2007. Energy Subsidies: Their Magnitude, How They Affect Energy Investment and Greenhouse Gas Emissions, and Prospects for Reform. Bonn: UNFCCC Financial and Technical Support Programme.
Morra Imas, Linda G., and Ray C. Rist. 2009. *The Road to Results: Designing and Conducting Effective Development Evaluations.* Washington DC: World Bank Publications.

O'Connell, Deborah. 2014. "Evaluating Sustainability." https://www.betterevaluation.org/en/blog/evaluating_sustainability.
OECD (Organisation for Economic Co-operation and Development). 2010. Evaluation in Development Agencies. Better Aid. Paris: OECD.
Okun, Arthur M., William Fellner, and Michael Wachter. 1975. "Inflation: Its Mechanics and Welfare Costs." *Brookings Papers on Economic Activity* 1975 (2):351–401.
Ostry, Jonathan David, Andrew Berg, and Charalambos G. Tsangarides. 2014. *Redistribution, Inequality, and Growth*. Washington DC: IMF.
OVE (Office of Evaluation and Oversight). 2015. Evaluating Climate Change. Washington DC: Inter-American Development Bank.
Petri, Peter, and Vinod Thomas. 2013. Development Imperatives for the Asian Century. *ADB Economics Working Paper no. 360*. Manila: ADB.
Picciotto, Robert. 2013. "Evaluation Independence in Organizations." *Journal of MultiDisciplinary Evaluation* 9 (20):18–32.
Piketty, Thomas. 2014. *Capital in the 21st Century*. Cambridge, MA: The Belknap Press of Harvard University Press.
Pradhan, Prajal, Luís Costa, Diego Rybski, Wolfgang Lucht, and Jürgen P. Kropp. 2017. "A Systematic Study of Sustainable Development Goal (SDG) Interactions." *Earth's Future* 5 (11):1169–1179.
Ravallion, Martin. 1997. "Can High-Inequality Developing Countries Escape Absolute Poverty?" *Economics Letters* 56 (1):51–57.
Ravallion, Martin. 2009. "Evaluation in the Practice of Development." *World Bank Research Observer* 24 (1):29–54.
Ravallion, Martin. 2013. How Long Will It Take to Lift One Billion People Out of Poverty? *Policy Research Working Paper no. 6325*.
Sabet, Shayda Mae, and Annette N. Brown. 2018. "Is Impact Evaluation Still on the Rise? The New Trends in 2010–2015." *Journal of Development Effectiveness* 10 (3):291–304.
Sabot, Richard, David Ross, and Nancy Birdsall. 2016. Inequality and Growth Reconsidered: Lessons from East Asia. *Working Papers id:8848, eSocialSciences*.
Sachs, Jeffrey D. 2012. "From Millennium Development Goals to Sustainable Development Goals." *Lancet* 379 (9832):2206–2211.
Shaker, Richard Ross. 2015. "The Spatial Distribution of Development in Europe and its Underlying Sustainability Correlations." *Applied Geography* 63:304–314.
Slough, Tara, and Johannes Urpelainen. 2018. Public Policy Under Limited State Capacity: Evidence from Deforestation Control in the Brazilian Amazon. *mimeo*.
Stern, Nicholas Herbert. 2006. *The Economics of Climate Change: The Stern Review* Cambridge: Cambridge University Press.
Takahashi, Ryo, and Yasuyuki Todo. 2012. "Impact of Community-Based Forest Management on Forest Protection: Evidence from an Aid-Funded Project in Ethiopia." *Environmental management* 50(3): 396–404.

Tarsilla, Michele. 2009. "Theorists' Theories of Evaluation: A Conversation with Jennifer Greene." *Journal of MultiDisciplinary Evaluation* 6 (13):209–219.
Thomas, Vinod, Mansoor Dailimi, Ashok Dhareshwar, Daniel Kaufmann, Nalin Kishor, Ramón López, and Yan Wang. 2000. *The Quality of Growth*. Washington DC: World Bank.
Thomas, Vinod, and Xubei Luo. 2012. *Multilateral Banks and the Development Process: Vital Links in the Results Chain*. New Brunswick, NJ: Transaction Publishers.
Tolley, George S. Graves, Philip E., and John L. Gardner. 1979. *Urban Growth Policy in a Market Economy*. New York: Academic Press.
United Nations. 2015. The Millennium Development Goals Report 2015. New York: United Nations.
United Nations. 2016. Global Sustainable Development Report: Briefs 2015. Geneva: United Nations.
United Nations. 2018. The Sustainable Development Goals Report 2018. New York: United Nations.
World Bank. 1991. World Development Report 1991: The Challenge of Development. Washington DC: World Bank.
World Bank. 2006. Development and the Next Generation. World Development Report 2007. Washington DC: World Bank.
Yamada, Hiroyuki, Yasuyuki Sawada, and Xubei Luo. 2013. "Why is Absenteeism Low among Public Health Workers in Lao PDR?" *Journal of Development Studies* 49 (1):125–133.

Open Access This chapter is licensed under the terms of the Creative Commons Attribution 4.0 International License (http://creativecommons.org/licenses/by/4.0/), which permits use, sharing, adaptation, distribution and reproduction in any medium or format, as long as you give appropriate credit to the original author(s) and the source, provide a link to the Creative Commons licence and indicate if changes were made.

The images or other third party material in this chapter are included in the chapter's Creative Commons licence, unless indicated otherwise in a credit line to the material. If material is not included in the chapter's Creative Commons licence and your intended use is not permitted by statutory regulation or exceeds the permitted use, you will need to obtain permission directly from the copyright holder.

CHAPTER 2

The Spectrum of Impact Evaluations

Abstract This chapter underscores the importance of causal attribution and takes the readers through various impact evaluation methodologies that enable evaluators to measure the causal impact of policies. Using case studies, it highlights important assumptions, advantages, and disadvantages of each methodology to give readers a sense of how these techniques can be applied to issues of sustainable development.

Keywords Attribution • Causal analysis • Counterfactual • Case studies

Policy-makers are increasingly seeking answers to the question of what works and what does not in addressing issues of making development more sustainable. Crucial to answering these questions is the ability to show, to the extent possible, attribution, that a change in the outcome, say a decrease in air or water pollution, is causally linked to a policy under consideration.

Conventional thinking is that the scale of issues such as climate change and income inequality is too large for them to lend themselves naturally to impact evaluations (IEs). Cameron, Mishra, and Brown (2016) did a systematic review of published IEs of international development interventions. Out of 2259 studies they reviewed (between 1981 and 2012, though the number of studies increased significantly after 2008), 1476 were on health, nutrition, and population; 521 were on education; and 341 were on social protection. Only 14 were on energy and 124 on environment and disaster management. The systematic review also finds that over the years there is stagnation on rigorous IEs on economic policy, energy, transportation, and urban development.

Given the complex nature of issues of sustainability, skepticism about their suitability for evaluation is justified. How can we randomly assign deforestation or air pollution to treatment and control areas? How can policy-makers experimentally roll out policies that are aimed at bridging rural-urban gaps? Is it politically feasible to provide social mobility opportunities to some but not to others? In this chapter, we discuss using real-life policy case studies how econometric experimental and quasi-experimental IE methods can be extended to overcome practical and empirical challenges.

Why Impact Evaluation?

Public policy-makers are guided by goals, objectives, and indicators. IE not only assesses whether goals were reached but also helps to understand the mechanism by which the impacts were generated. The shift toward evidence-based policy-making calls for a good understanding of what IE can and cannot do as well as how it can be designed, applied, and replicated. The core objective of IE is to assess how much of an impact can be attributed or causally linked to a specific project, program, policy, or even a shock such as a climate-related natural disaster.

The application of IE is not limited to small-scale and targeted projects. IE tools offer flexibility to evaluate targeted projects such as the impact of change in classroom pedagogy in public schools on children's learning outcomes, as well as to evaluate large-scale, national-level programs and policies such as compulsory education or subsidies for the education of girls. It thus has the capacity to assess the impact of interventions related to an array of issues included in the SDGs such as poverty alleviation, social inclusion, and environmental stewardship.

To illustrate the practical and empirical challenges in conducting IEs of sustainable development policies, let us consider, as an example, the policy to control illegal deforestation in Brazil, which is directly related to SDG 12 (climate action). We know that deforestation activities tend to be concentrated in areas with high levels of forest resources and low levels of governance and monitoring. Therefore, policies to control deforestation are often geographically targeted. One such policy is the Priority Municipality (PM) program introduced in the Brazilian Amazon in 2007 that instated rigorous monitoring in areas that experienced extensive illegal deforestation (Slough and Urpelainen 2018). While this is the typical policy response, assessing the success of such a policy in curbing deforestation can be challenging.

An obvious challenge for evaluators is that the PM program does not have control over the movement of extractors engaged in illicit deforestation from the priority areas to other areas where there is no restriction. It is thus possible that extractors who previously practiced illegal logging in priority areas decided to move elsewhere and continue their activities after the program was introduced. If an evaluation only accounts for the changes in deforestation within the priority areas and fails to consider the negative spillovers in other areas, the results may suggest a significant decrease in deforestation rates. This might lead to the conclusion that the program achieved its objectives although there may be serious negative spillovers in other areas.

The question IE should ask therefore is "What is the *treatment effect* of a program on an outcome?" Answering this question requires a good understanding of causal inference and the spectrum of available evaluation methods so that the most suitable one can be chosen.

Causal Inference and Counterfactuals

In seeking answers to questions about the effect of an intervention, the challenge is to establish causality between a program and an outcome. Econometric IE is a tool that helps us empirically establish causality by measuring the differences (Δ) in outcome (Y) of the program participants (T) and outcome of the nonparticipants (C), given by the formula:

$$\Delta = (Y|T) - (Y|C)$$

To further illustrate the complexity in establishing causality, let us look at another example. Investing in rural infrastructure such as electrification and roads is considered vital to reducing rural-urban inequality and promoting inclusive growth (UN 2016). Chen, Chindarkar, and Xiao (2019) examine the causal effect of an electrification upgrading program on improvements in rural health systems including health services utilization, health information, and health facilities. In 2003 the state government of Gujarat, India, launched the Jyotigram Yojana (JGY), which provides 24-hour, high-quality electricity to rural areas. Stable electricity supply is an enabler of universal access to health care as it mediates health services utilization such as child immunization and ante-natal care, access to health information through electronic media, and functioning of health equipment in rural health centers (Chen et al. 2019; WHO 2014).

Therefore, improving the quality of electricity supply can be expected to improve rural health systems. This would effectively result in greater health equity as the health gap between rural and urban households is narrowed. Considering just one of the outcome indicators—child immunization—the formula indicates that the gap between the immunization rate of children from households that reside in villages that were electrified under the program $(Y|T)$ and the immunization rate of children from households that reside in villages which remained unelectrified $(Y|C)$ is the effect caused by program (Δ). An important question is whether the households in electrified and unelectrified villages are comparable.

In order to draw a causal inference, the observed outcome of the treatment group—or the individuals or households affected by the program—needs to be compared with the potential outcome of the group had they not been exposed to the program. This is referred to as the counterfactual outcome. The difference between the actual outcome and counterfactual outcome can be attributed to the program because in this scenario the two groups are identical in expectation except for their treatment status.[1] In other words, we expect the two groups to be identical, on average, in the absence of the program. In reality, however, the counterfactual outcome is not observed. It is not possible to observe the immunization rate of the same children with and without electrification simultaneously. The counterfactual thus needs to be estimated, and this is where econometric tools come in handy.

Estimating the Counterfactual

The identification of program impact requires generating two groups that are statistically identical in expectation in the absence of the program: one group affected by the program, called the treatment group, and a group not affected by the program, called the control group. Comparing outcomes of these groups ensures that the difference between the outcomes of the treatment group and the control group is due to the program. The challenge in identifying the causal impact is to find a valid control group.

In the case of rural electrification, the first thing that one would think of is to compare the immunization rate before the treated households were exposed to the program and after the program was implemented:

$$\Delta = (Y_{t1} | T) - (Y_{t0} | T)$$

As illustrated in Fig. 2.1, the immunization rate before the intervention, which is an estimate of the counterfactual, is Y_{t0}, and the rate after the intervention is Y_{t1}. The before-and-after comparison seems to suggest that the intervention increased the immunization rate by $Y_{t1} - Y_{t0}$.

However, consider the case where the majority of rural households suffered from a drought. A decline in income from the crop damage could have discouraged parents from investing in their children's health. Then the outcome in year 1 in the absence of intervention would likely be lower than Y_{t0}. In this case, the actual impact of the program might be $Y_{t1} - Y_{t1'}$, which is larger than $Y_{t1} - Y_{t0}$. A failure to account for the effect of drought will result in underestimating the impact.

On the other hand, other factors might positively affect the health outcomes of children over time such as an increase in household income or

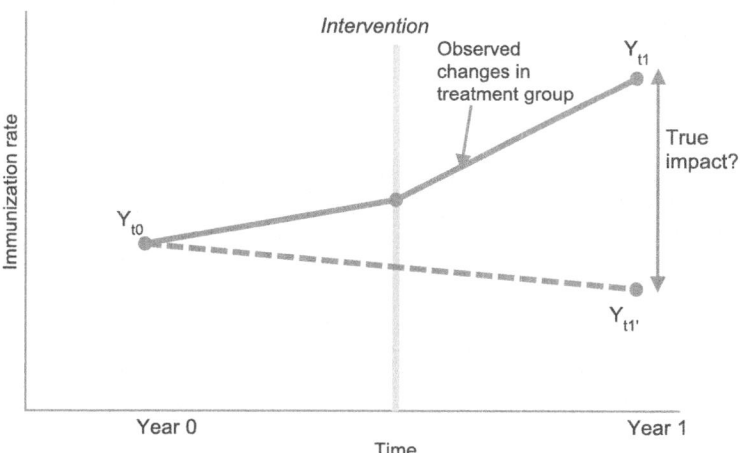

Fig. 2.1 Before-and-after comparison. (Source: Authors' illustration)

increase in health budgets allocated to the state. Ignoring these factors that might lead to an increase in the child immunization rate could result in overestimation of the impact of the program. The same considerations hold for other types of interventions, such as education, vocational training, and micro credit. The baseline outcome can hardly serve as an accurate measure of the counterfactual.

With such situations in mind, another method of counterfactual estimation uses a group that is not exposed to a program. As shown in Fig. 2.2, the gap in observed outcomes of the treatment group and the control group is $Y_{t1} - Y_{c1}$. This estimates the impact of the program only if we can assume that the *changes* in the immunization rates caused by the electrification program would not be different for the two groups. In many cases, the program is targeted toward areas or groups of people who are in need of the program. If the households in target areas of rural electrification program are different from households in non-target areas, say in terms of their socioeconomic conditions, then we are essentially comparing apples to oranges.

The effect of an intervention is then likely to be different for the treatment group and the control group. If the counterfactual of the treatment group were observed, then we would know that the real impact is observed

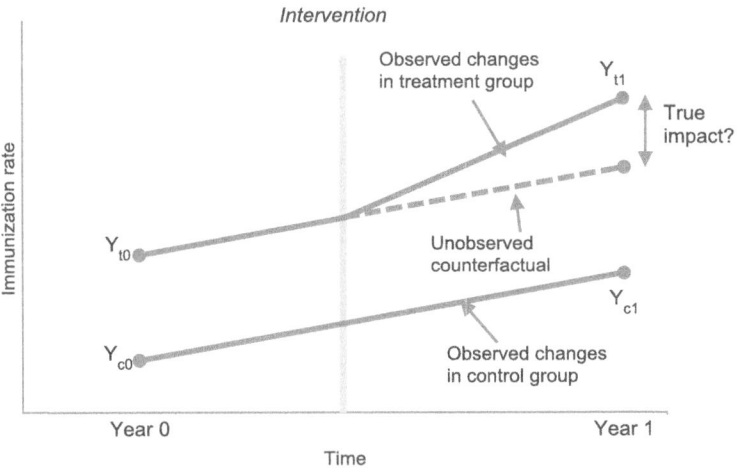

Fig. 2.2 With-and-without comparison. (Source: Authors' illustration)

changes in treatment group ($Y_{t1} - Y_{t0}$) less observed changes in control group ($Y_{c1} - Y_{c0}$). In other words, the change in outcome of the control group is the change that would have occurred without any policy intervention and therefore that part cannot be attributed to the intervention.

Another important factor that could taint the estimate of counterfactual is the selection mechanism. Information on the electrification program could attract some households to move from villages with poor electricity supply to the target villages. Those households might also have higher health awareness than those who stay in non-targeted villages. Naturally, the effect of the program is larger for such households compared to others, which again leads to a bias in the program's impact estimation.

Selection bias occurs when the program participation decision is correlated with unobserved factors. This is a serious concern in various types of interventions. In a conditional cash transfer program where the cash is provided on the condition of children being in school, it is highly likely that parents with higher motivation to send children to school, which is typically unobserved, will participate in the program. Depending on the selection mechanism, a simple with-and-without comparison could bias the estimated impact of the program.

Establishing a Theory of Change

Before applying either experimental or quasi-experimental methods, evaluators must lay out the theory of change, which is the causal logic of how and why a particular program is likely to achieve the intended outcomes. It is guided by existing theoretical and empirical literature and helps to build appropriate hypotheses for the IE. For instance, the theory of change underlying improved rural electrification and child immunization may be that electrification provides better vaccine storage facilities. There might also be positive spillovers from improvements in other aspects of the health system such as improved health facilities and increased access to health information.

In the IE, the econometric analysis would examine the effects of improved electricity access on child immunization rates as well as mediating factors such as health facilities and health information.

Internal Validity and External Validity

IE seeks answers to the causal effect question, and its usefulness greatly depends on its validity, both internal and external. For an evaluation to have internal validity, the outcome of the control group needs to be indeed a valid counterfactual and the estimated impact needs to be solely attributed to the program. In some cases, explanatory variables may not be well specified (referred to as omitted variables) or accurately measured (measurement error). There might be issues related to program assignment (imperfect compliance with the treatment or correlation between treatment assignment and outcome). Other factors such as attrition and externalities also put internal validity at risk. Any of these might cause low internal validity, undermining the inference of causality.

External validity means that results are applicable or generalizable to different populations, contexts, and outcomes. The threats to external validity essentially concern important interactions between the treatment and individual characteristics, location, or time (Meyer 1995). The less the likelihood of violating external validity, the more confident policy-makers can be in applying the impact evaluation learning to populations beyond the one under examination, or to other contexts.

While strong validity improves the quality of IE, incorporating IE in policy design a priori could help minimize threats to internal and external validity. Prospective IE can be incorporated in the process of policy design so that valid counterfactuals and data are available in the future.

Random Assignment

It is not possible to avoid all threats to internal and external validity, but there is a tool to help deal with it: random assignment. Often referred to as randomized controlled trials (RCTs), random assignment is increasingly used in economics and other social sciences. RCTs give every eligible unit an equal probability of being selected into a program. Such a selection mechanism not only generates a valid counterfactual, but it is also transparent and accountable. RCTs are often viewed as the most credible approach to establishing causality because they require few statistical assumptions and analysis can be done using simple econometric methods.

Random assignment of treatment and control groups produces two comparable groups when sample size is large. This is based on the property called the law of large numbers (LLN). The LLN states that a sample average will approximate the average of the population from which it is drawn as the sample size grows larger. The gap between averages of two groups can then be interpreted as the unbiased estimator of the average treatment effect (ATE). This can be expressed as

$$Y_i = \alpha + \beta T_i + \varepsilon_i$$

where T_i is the treatment status dummy that equals 1 if a randomly selected unit, i, is treated, and 0 otherwise. Random assignment ensures that T and i are independent and the estimated treatment effect $\hat{\beta}_{OLS}$ is unbiased.

Sampling and Validity Issues in Randomization

In practice, it is not simple to generate two groups that are the same except for the treatment status. Random assignment is commonly conducted in two steps. The first step is to randomly select a sample of potential participants from the eligible population. The second step is to randomly select units to be assigned to treatment and control groups. Each step ensures external and internal validity, as in Fig. 2.3.

Although RCTs seem to be a solution for establishing validity, some have pointed out that in reality they might be compromised (Deaton and Cartwright 2016; Ravallion 2018). Internal validity of estimates could be

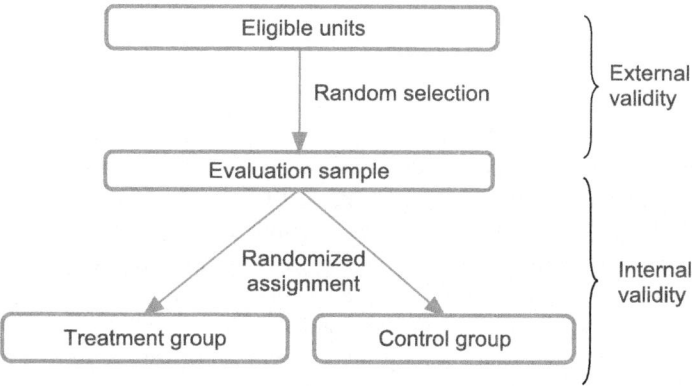

Fig. 2.3 Random sampling and randomized assignment of treatment. (Source: Authors' illustration)

put at risk when compliance with treatment assignment is not perfect, there are externalities, or randomization is conditional on observed variables. Most programs aim to reach all members of the randomly assigned treatment group. In many cases, however, full compliance with the treatment assignment may not be achieved. This could be due to the behavior of both the treatment and control groups.

Take, for example, a vocational training program offered to randomly selected schools. Some in the treatment schools who are offered a free training course may not be motivated to take up the program. On the other hand, some in the control group may decide to transfer to a school in the treatment group. These behaviors change the original treatment assignment status and contaminate the randomized design.

A second threat to internal validity is externalities. Most social science RCTs are conducted in the field, where externalities are often generated, and not in a laboratory. In the vocational training program example, consider a case where there are two friends, one assigned to the treatment group and the other assigned to the control group. It is conceivable that the one in the treatment group performs well owing to the training program and passes on information to the friend in the control group, who, because of the information received, also performs well.

A third threat is imperfect randomization. Randomization is often conditional on a set of observed variables. The assumption that, conditional on the observed variables, the potential outcomes of treatment and control groups are identical in expectation could eliminate the selection bias. For this assumption to hold, the set of observed variables needs to include all the relevant variables that account for the differences between the two groups. Incomplete data on observables could result in selection bias owing to omitted variables. It is, therefore, important to carefully select the variables according to the setting and purpose of the program.

While randomization can eliminate selection bias to a large extent, it does not guarantee that findings from an RCT in one context will necessarily hold in others. Duflo, Glennerster, and Kremer (2007) discuss that there are three major factors to consider in examining the external validity of RCT results. First is careful design and documentation of the intervention. While an RCT can be well designed and successfully implemented as a pilot or on a small scale, scaling up and replication can present a bigger challenge, particularly when the evaluation design is not clearly documented. Procedures must be put in place to record the study design and implementation processes so that policy-makers can use them for program expansion and replication.

A second factor is whether findings from an RCT on one sample can be generalized to the population. As previously discussed, external validity can be strengthened by randomly selecting the sample from the eligible population in the first step and randomly assigning the treatment and control groups in the second step. However, in practice, sampling may not be random and therefore not representative of the population. In some cases, the eligible sample may be selected because it is convenient. The sampling decision is often guided by the availability and approval of study partners such as nongovernment organizations or local governments. This severely constrains the generalizability of RCT findings to samples beyond the one being studied.

A third factor is how the effect of the program would differ if the treatment were slightly different. In a conditional cash transfer program, what would happen if the amount offered were increased? Would the results change if the age of eligibility were lowered? Conducting RCTs with multiple treatment arms could offer insights into what works and what does not and how the program could be tweaked. However, this increases the design and implementation complexity. A sound approach is to have an appropriate theoretical framework to judge which treatment arms are important.

Finally, RCTs may not always yield estimators that are more unbiased relative to observational or nonrandom studies. The first reason for this is linked to issues of external validity and choice of samples for RCTs (Ravallion 2018). The second reason pertains to the variance of errors in estimates from RCTs compared to observational studies. Ravallion (2018) argues that, despite the bias, the variance of errors from observational studies that use large sample sizes could be low enough to assure that they are closer to the true population parameter. In contrast, despite the lack of bias, the small and selective samples that are often used in practice for conducting RCTs may yield estimates that are further from the true population parameter.

Which Treatment Parameters Are of Interest?

Deaton and Cartwright (2016) suggest there are three alternatives when internal validity is violated and an ideal ATE, β, cannot be calculated. First is to calculate the difference in means between those who, regardless of their assignment status, received the treatment, β_1, and those who did not. This is called the average treatment effect on the treated (ATT).

Second is to estimate what is called "intent to treat" (ITT), β_2, which is the difference between the average outcome of those who were intended to be in the treatment group and those who were intended to be in the control group, according to the original treatment assignment and regardless of whether they complied. The ITT estimate will be different from β unless there is perfect compliance. Perfect compliance is often violated in field experiments, and those who do not comply with their assignment status tend to have different characteristics compared to those who comply, making β_2 a parameter of interest.

Third is an estimator, β_3, called the local average treatment effect (LATE). In many cases, the program is not directly offered to all individuals. Rather only information on the program is randomized, and individuals select themselves into the program based on the information. Such experimental design is common in social programs offering vocational training. LATE estimates the program effects for the subgroup that complies (p), and it is calculated as $\beta_3 = \beta_2 / p$. In particular, it only accounts for those whose treatment status was induced by the randomized program information. In other words, it is the average causal effect for those who participated in the vocational training program only because they were offered information without which they would not have participated.

These three estimators are average over different populations; therefore, they are different without additional assumptions on the heterogeneity of treatment effects (Deaton and Cartwright 2016). In general, it is natural to assume there are different characteristics for those who comply with the treatment assignment and those who do not. For instance, those who are offered information but decide not to participate in the vocational training program may already have high skills and not feel the need for further training. Those who participate even if they are not offered information may have higher motivation and may learn more from the same training than participants in the treatment group. Given that treatment effects are often heterogeneous, it is vital to be clear about what is being evaluated and which treatment parameter is being estimated.

Need for Baseline Information

Since the treatment assignment may not be completely random even in RCTs, a good practice is to conduct baseline surveys to examine initial conditions as well as their interactions with the impact of the program. Baseline surveys are crucial in conducting balance checks. Balance checks enable evaluators to statistically judge whether the treatment and control groups are similar on average before the intervention is introduced. These can be performed using simple hypothesis tests of difference in two sample means. The expectation is that there should be no systematic differences on average in the observed characteristics of the treatment and control groups. This strengthens both the internal and external validity of the findings from the RCT.

Additional balance checks can be performed in case of attrition, where units assigned to the treatment or control group drop out of the experimental study. Here we would compare the balance between the treatment and control groups before and after attrition. Again, the expectation is that there are no differences between the treatment and control groups post attrition, meaning attrition was not systematic and therefore should not be a validity concern.

Quasi-experimental Methods

Can policy-makers randomly select where to construct a new road, build irrigation systems, or supply electricity? RCTs have many advantages; however, public policy implementation rarely follows experimental design, as it may not fit with program objectives, be costly, or be politically unfeasible or unethical. In circumstances where randomization is not feasible, it is possible for policy analysts to exploit natural experiments or quasi-experiments that offer opportunities to select a control group that was excluded from the program but shares similar characteristics with the treated group.

Various econometric tools are available to identify causal effects using quasi-experimental methods. Each method comes with a different set of assumptions and data requirements that need to be considered carefully. Each has its advantages as well as limitations. In this section, we will discuss four commonly used quasi-experimental methods—difference-in-differences, regression discontinuity, instrumental variables, and propensity score matching.

Difference-in-Differences

Interventions to tackle environmental issues often adopt geographical targeting as policy needs to prioritize areas with more severe environmental deterioration. The study of deforestation policy in the Brazilian Amazon is a case when randomized policy implementation is not possible because, by its very nature, priority areas are located where deforestation activities are more extensive (Slough and Urpelainen 2018). This program which introduced rigorous monitoring in areas with extensive deforestation could generate a displacement of deforestation to neighboring areas if there is limited state capacity to properly implement the program.

Evaluations if designed properly can help examine the effect of policies beyond the targeted areas. To create a natural experiment setting, that is, assignment of priority areas, the study uses changes in priority areas over time according to changes in deforestation rates. The study combines information on priority areas designated by the government and from satellite monitoring data to identify forest clearing. It then exploits the variation in designation of priority areas over time to evaluate the impact on deforestation in the target areas as well as neighboring areas using the difference-in-differences (DID) method.

A simple before-and-after or with-and-without comparison would not give an accurate causal estimate of such a program. The quasi-experimental setting always leaves concern about nonrandom program implementation. If both pre-program and post-program data are available for both treatment and control groups, a method called DID is one way to eliminate the bias. DID makes use of these data to obtain a valid counterfactual to estimate the effect of an intervention or a program by comparing the average change in outcome over time between the treatment group and control group.

Using the example of deforestation policy, average change in the probability of a deforestation event (the outcome variable of interest) in the priority area and control area is illustrated in Fig. 2.4. Before the program, the average probability of a deforestation event for the treatment group is A, which is higher than the average probability for the control group, C. The average change from year 0 to year 1 in the control area is the counterfactual for the priority area. This means that in the absence of the program, the priority area would follow the same trend as the control area and reach E in year 1. This decrease from A to E needs to be subtracted from the actual change in the treatment group, from A to B.

Therefore, as summarized in Table 2.1, the impact of the program is calculated as

$$\text{DID} = (B-A)-(D-C) = (0.76-0.86)-(0.68-0.70) = -0.08$$

THE SPECTRUM OF IMPACT EVALUATIONS 39

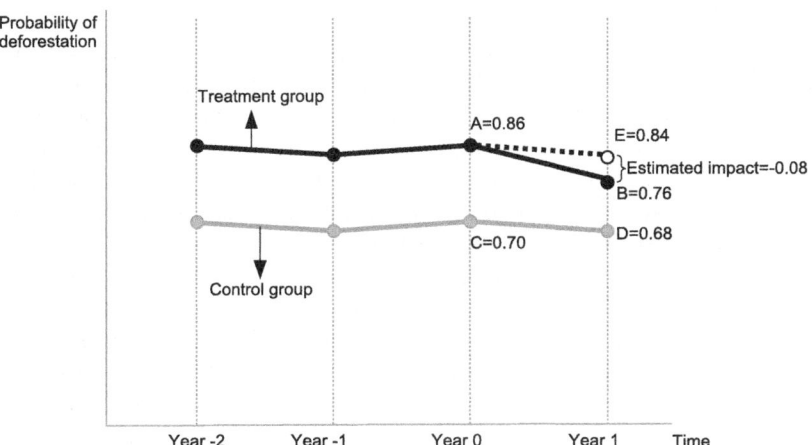

Fig. 2.4 DID applied to deforestation policy. (Source: Authors' illustration)

Table 2.1 Calculating the impact in DID method

	After	Before	Difference
Treatment group	B	A	B – A
	0.76	0.86	–0.10
Control group	D	C	D – C
	0.68	0.70	–0.02
Difference	B – D	A – C	(B – A) – (D – C)
	0.08	0.16	–0.08

Source: Authors' illustration

As DID averages the treatment effect over the entire treatment and control population, the resulting estimate is the ATE. Econometrically, estimation of the ATE using DID is done using the following regression model:

$$Y_{it} = \beta_0 + \beta_1 TREAT_i + \beta_2 POST_t + \beta_3 TREAT_i * POST_t + \varepsilon_{it}$$

where Y_{it} is the outcome variable of interest. $TREAT_i$ equals 1 if treated and equals 0 if not treated. $POST_t$ equals 1 if post-program period and equals 0 if preprogram period. β_3 gives the DID estimate.

The DID Parallel-Trends Assumption

DID provides an unbiased estimate of the treatment effect under the parallel-trends (or equal-trends) assumption. The parallel-trends assumption is that in the absence of the program, the difference in the outcome over time for the treatment and control groups would follow the same trend or the outcomes would move in tandem. It is important to note that the assumption does not state that in the absence of the program the outcome for the treatment and control groups would be the same. Rather, it assumes that the outcomes, although different, follow the same trend.

Although there is no formal statistical test to prove that both groups follow the same trends in the absence of a program, there are several ways to check the validity of this assumption. First is to graphically compare the trends in outcome using several periods of preprogram data. If the pre-policy trends are the same for the treatment and control groups, then it is safe to assume that they follow parallel trends. Figure 2.4 shows that control and treatment groups follow the same trends before the program implementation. This gives more confidence in assuming that the two groups would follow the same trends after year 0 if not for the program.

A second way to test the validity of the parallel-trends assumption is to perform what is known as a placebo or a falsification test using a different slice of time, a different sample, or a different outcome variable. The idea is that the program should have no impact on this differently chosen time, sample, or variable. If we do find significant effects, then there might be some unaccounted-for or unobserved factors outside of the program that caused the changes in outcome. Slough and Urpelainen (2018) use the 12-month time period prior to actual priority area assignment and run their DID model only on pre-program data. The hypothesis is that there should be no significant reduction in the probability of deforestation during this period. Their placebo test results support this hypothesis, and therefore the parallel-trends assumption holds.

Advantages and Limitations of DID

In many quasi-experimental programs, the treatment assignment rules are not as clear as in experiments. The advantage of DID is that it controls for unobserved as well as observed characteristics that affect participation in the program as long as the characteristics are time invariant. Many observed characteristics, such as geographic and climatic conditions, or unobserved characteristics, such as a culture of conservation, are likely to be constant over time. DID's biggest limitation, however, is that it does not control for time-varying unobserved characteristics, like the ability and motivation of local government personnel in implementing deforestation policy. If different personnel are in charge at different points in time, their ability and motivation to implement the policy with stringency is likely to be time varying. Therefore, we might still estimate slightly biased treatment effects.

Regression Discontinuity Design (RDD)

Often public policies follow eligibility criteria for targeting purposes. Common examples of these are pension programs, which impose an age eligibility criterion, or poverty-alleviation programs, which impose a minimum-income criterion. These criteria can be exploited to create comparable treatment and control groups and to evaluate large-scale programs. The example of an evaluation by Chen et al. (2013) of energy policy in China can help illustrate this. The study applies a quasi-experimental method called regression discontinuity design (RDD) to evaluate an energy program that provides coal for winter heating in Northern China.

RDD can be used to evaluate programs that have a continuous eligibility index with a clearly defined eligibility threshold or cutoff. The observations close to the cutoff are divided into the eligible (treatment) and non-eligible (control) groups, and their outcomes are compared in order to estimate the local average treatment effect. As RDD restricts the treatment and control groups only to a certain bandwidth around the cutoff to ensure that they are similar on average, the treatment effect cannot be generalized to the entire population. We are therefore only able to estimate the LATE.

During the period of central planning (1950–1980), the Chinese government provided free coal for winter heating to homes and businesses as a basic right in Northern China. Such coal combustion releases harmful air pollutants that are known to adversely affect human health. Owing to budgetary limitations, the free provision was restricted only to areas north of the Huai River (shown in Fig. 2.5). This created a quasi-experimental opportunity to compare the cardiorespiratory mortality rates and life expectancy of the treatment group, residing just north of the river that received free coal, and the control group, residing just south of the river that did not receive free coal. Here the distance from the river is the continuous eligibility index, and the river itself is the spatial cutoff point. As the two groups reside within close proximity to the river, they are assumed

Fig. 2.5 Cities to the north and south of the Huai River. (Source: Chen et al. 2013)

to be similar in all important aspects except for the amount of pollutants they were exposed to.

The RDD treatment effect can be estimated using the following linear regression model:

$$Y_i = \beta_0 + \beta_1 x_i + \rho w_i + \varepsilon_i$$

where Y_i is the value of the outcome for unit i, in this case, life expectancy at birth; x_i is the continuous eligibility index, in this case, the degrees north of the Huai River; w_i is the dummy variable that indicates whether the unit is in the treatment or the control group, in this case, 1 for locations north of the Huai River and 0 otherwise.

The study finds a striking decline in life expectancy north of the Huai River. Figure 2.6 indicates that average life expectancy at birth reduced by

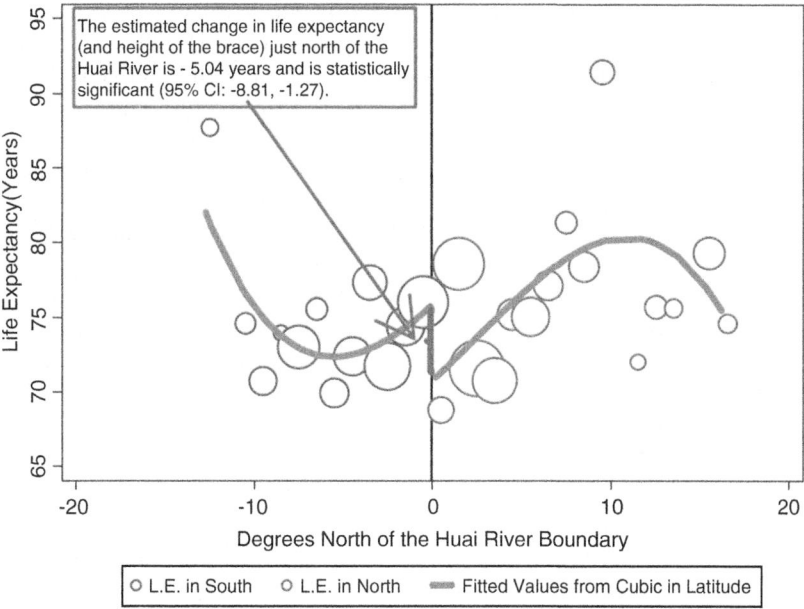

Fig. 2.6 Fitted values from RDD estimation. (Source: Chen et al. 2013)

almost five years for those living just north of the river owing to increased exposure to air pollution.

This study demonstrates the difficult trade-off between economic growth, public health, and environmental quality that many growing economies face today. The level of pollutants at the time of the study could be used as a reference for cities in developing countries such as Brazil and India where pollution is a serious issue. Though certain adjustments are required in applying the findings in different countries or different contexts, the insights obtained in such evaluations are useful in controlling or avoiding interventions that might have negative consequences on the environment.

Advantages and Limitations of the Regression Discontinuity Design (RDD)
The RDD method exploits the opportunities naturally generated by the program eligibility criteria and allows unbiased estimates of the treatment

effect. An advantage of the RDD method is that it does not require any eligible units to be untreated for the purposes of the IE. The treatment effect, however, is valid only for the units around the eligibility cutoff. In other words, the estimated treatment effect is the LATE. Therefore, an important limitation of RDD is that the estimated effects may not always be generalized to units whose eligibility scores are far from the cutoff point.

A further challenge arises when the enforcement of eligibility is not clear-cut or "sharp" but is "fuzzy." This means that not all eligible units may be affected by the program, and some ineligible units might be affected. If the compliance with the eligibility criteria is "fuzzy," the eligibility score can be replaced with a probability of participating, and the estimated treatment effect is the difference around a neighborhood of the cutoff score.[2]

The statistical power of analysis presents another challenge that arises because the RDD method only estimates impact around the cutoff. This restricts the number of observations used to estimate the impact, which lowers the statistical power of analysis. The bandwidth around the cutoff point needs to be determined so as to include a sufficient number of observations while maintaining the balance of important characteristics to make the treatment and control groups comparable.

Finally, problems also arise when it is possible for participants to manipulate eligibility criteria. For instance, if corruption is high, it may be possible for people to provide fake documents to make them eligible for the program. This contaminates the quasi-experimental features of RDD and produces biased estimates. A simple way to identify manipulation is to plot a histogram of eligible units along the continuous eligibility criteria. The appearance of far too many units clustered just above the eligibility criteria might indicate potential manipulation, and policy analysts will need to dig further into how the program was implemented on the ground.

Instrumental Variables (IV)

Another important quasi-experimental method is called instrumental variables (IV). As discussed previously, there might be a systematic correlation between program participation and unobserved characteristics of the participants, in what is often referred to as endogeneity. Endogeneity may arise if participants self-select themselves into a program or they do not comply with randomized experimental design or program eligibility criteria. IVs allow us to address such issues of endogeneity.

Let us understand IV using IE of the Moving to Opportunity (MTO) program, an experimental housing-mobility program introduced in 1994 in several cities in the United States. This program was motivated by the fact that there is significant geographical disparity in social and economic status and was implemented to examine whether moving from a high-poverty neighborhood to a low-poverty neighborhood improves social and economic prospects of low-income families. Under MTO, eligible families were randomly assigned housing vouchers by the US Department of Housing and Urban Development to move from poorer neighborhoods to better-off neighborhoods. They were also provided counseling services to adjust to the new neighborhoods. The control group families did not receive any vouchers.

However, they continued to receive other government assistance they were eligible for. As discussed in the randomization section, in reality, perfect compliance with the randomized treatment assignment is rare. In the case of MTO as well, not all families who were offered the housing vouchers actually took them up. The evaluation of MTO conducted by Chetty, Hendren, and Katz (2016) applies IV to address this imperfect compliance in voucher take-up.

Without full compliance, the estimated treatment effect is either that of offering a program (ITT), that of participating in the program (ATT), or limited to those who complied with the experimental design or program eligibility criteria (LATE). As previously discussed, the basic ATE estimation regression setup is expressed as follows:

$$Y_i = \alpha + \beta T_i + \varepsilon_i$$

When treatment assignment is not random in reality, treatment dummy T and the error term ε are systematically correlated, that is, $\text{cov}(T, \varepsilon) \neq 0$. The IV method aims to remove this correlation by isolating the variation in T that is uncorrelated with ε. For an instrumental variable, Z, to be valid, it must satisfy the following two conditions:

$$\text{cov}(Z, T) \neq 0 \text{ and } \text{cov}(Z, \varepsilon) = 0$$

The first condition is called relevance, and it shows that an IV is correlated with the treatment variable. The second, called exogeneity, shows that the IV is uncorrelated with the error term. Essentially, we rule out any direct effect of the IV on the outcome or any effect coming from unobserved or omitted variables. This is also known as exclusion restriction.

Chetty, Hendren, and Katz (2016) use the randomly assigned MTO treatment indicator as an IV (Z) for actual take-up of housing vouchers. As the random treatment indicator is correlated with treatment assignment and uncorrelated with the error term, it satisfies the IV validity conditions. The exclusion restriction is that the MTO voucher offers affect the outcomes only through the actual use of the voucher. They use a two-stage least-squares (2SLS) regression that is composed of two regressions.[3] The first stage regresses the voucher dummy variables on the random treatment indicator Z, additional covariates, and the error term, u_{1i}:

$$T_i = \pi_0 + \pi_1 Z_i + u_{1i}$$

Because Z_i is uncorrelated with u_{1i}, the estimate of π_0 and π_1 is uncorrelated with u_{1i}.

The second stage regresses the outcome variable on the predicted value of voucher take-up from the first stage with other covariates and the error term:

$$Y_i = \alpha_0 + \lambda \widehat{T}_i + u_i$$

Because $\widehat{T}_i$ is uncorrelated with u_i, we can now say that the correlation between the treatment variable and the error term is zero in the second stage. In other words, voucher take-up is no longer systematically correlated with the error term. From the 2SLS estimates, Chetty, Hendren, and Katz (2016) find that children who moved to better-off neighborhoods before the age of 13 years had better rates of college attendance, higher earnings, and lower rates of single parenthood as compared to children who did not get the opportunity to move. When applied to a broader context, programs such as MTO are likely to reduce intergenerational transmission of poverty and inequality.

IV is also useful in evaluating infrastructure programs, which are often targeted toward specific areas. In South Africa in 1993, where only one-third of the households had access to electricity, the government committed itself to universal electrification. By 2001, almost a quarter of households were newly connected to the grid due to mass rollout of electricity. Evaluating the causal effects of the intervention is not straightforward, as program implementation was not random.

To address the selection bias, Dinkelman (2011) uses an IV approach and analyzes the impact of access to grid electricity on employment growth in rural communities. Electrification implementation is instrumented using land gradient. Land gradient is an important determinant of implementation sequence as more time and resources might be required to connect communities residing in higher altitudes and therefore they might be connected to the grid later compared to communities residing on flat lands. The exclusion restriction of the study is that land gradient is unlikely to affect employment outcomes other than through electrification.

Moreover, IV is also suitable to evaluate the effect of good governance on economic growth, which often suffers from endogeneity because governance and economic growth affect each other simultaneously, that is, good governance can increase economic growth but at the same time economic growth can lead to improved governance. Mauro (1995) analyzes data from 70 countries with information on corruption, red tape, and efficiency of the judicial system. Among these institutional factors, he finds that corruption is the cause for lower private investment, which leads to lower economic growth. The IV used to address endogeneity is the index of ethnolinguistic fractionalization, which measures the probability that two persons drawn at random from a country's population will not belong to the same ethnolinguistic group.

The IV meets the two conditions of relevance and exogeneity—countries with higher fractionalization are expected to be more corrupt as bureaucrats may favor their own ethnolinguistic groups; and fractionalization is not expected to directly affect economic growth other than through its effect on institutional efficiency. Not only does the study identify the channel through which governance affects economic growth, but it also estimates the magnitude of the effects, which offers valuable insights into policy-making. For example, the findings suggest that if Bangladesh improves its integrity and efficiency of bureaucracy to the level of Uruguay, its investment rate would rise by almost 5 percentage points and its annual GDP growth rate would rise by over 0.5 percentage points.

Advantages and Limitations of IV

IV enables evaluators to obtain unbiased estimates of treatment effects even in the presence of imperfect compliance. A significant advantage of IV is that evaluators can apply the method even to post-program cross-sectional data. A drawback is that it is not always feasible to find a valid IV. Unless the IV satisfies the validity conditions, the estimates of the program effect will be biased. Since there is no statistical test for exclusion restriction, one has to draw upon theory and policy background to argue that the IV is truly exogenous. Only under a very specific condition of availability of multiple IVs can a statistical test for weak instruments be conducted.

Propensity Score Matching (PSM)

How can we evaluate a program if we do not have pre- and post-policy data, a clear eligibility criterion, or a valid IV? A quasi-experimental method available to us under such circumstances is propensity score matching (PSM). It can be applied when we only have post-program data. PSM constructs an artificial comparison group by selecting units from the untreated group that share similar observed characteristics with the treated units. As long as there is an untreated group, PSM does not require explicit treatment assignment rules. Another important feature of PSM, that it does not require one-to-one matching of all the relevant observed characteristics, has opened up opportunities for program evaluators to apply matching techniques in IE.

The first step in PSM is to compute the propensity score, which is the probability of being treated calculated using observed characteristics, including factors that influence treatment assignment as well as the outcome. This is done by running a probit or logit regression with the treatment dummy as the outcome variable and all relevant observed characteristics as covariates. The calculated predicted probabilities are then used to identify the treated and untreated units that have the same or extremely close propensity scores. Similar or close propensity scores imply that the treated and untreated units share the same characteristics. The matched treated and untreated units then form the treatment group and the (artificial) control group.

To further explain PSM, let us consider the study by Capuno and Garcia (2010) on the evaluation of a good governance program in the Philippines. The Good Governance and Local Development Project (GGLD) was established with the aim of institutionalizing a set of indicators to track the performance of local governments in the Philippines called the Governance for Local Development Index or Gofordev Index (GI). GGLD was first implemented in 12 local governments in the Bulacan and Davao del Norte provinces during 2001–2003. In eight out of twelve local governments, GI scores were generated and disseminated to the public to make them aware of the performance of their local governments. In the remaining local governments, the GI scores were generated but not disseminated.

GI assessed the Local Government Units (LGU) based on three performance domains: public service needs (access to and adequacy of basic services and the perceived effectiveness of the LGU in improving family welfare), expenditure prioritization (share of health, education, and other basic services in total fiscal outlays), and participatory development (functioning of the local consultative bodies and the public consultations at the village level). As citizens in LGUs with and without GI dissemination were not directly comparable, PSM was used to generate comparable treatment and control groups. The objective of the evaluation was to examine whether better knowledge of the performance of local government increased civic participation among citizens. The civic participation outcomes are dummies indicating membership in local organizations and participation in local projects.

The propensity scores are computed using a probit regression that controls for all possible relevant observed characteristics that determine the probability of knowledge of GI and also the outcomes. This can be written in the following regression form:

$$P(w_i = 1|X) = G(X\beta) \equiv p(X)$$

where w_i is the probability that the individual is in the treated LGU conditional on all the observed characteristics captured in the vector X. The propensity score is denoted by $p(X)$. In order to minimize selection bias, each individual in LGUs where GI scores were disseminated is matched with an individual in LGUs where the scores were not disseminated. This is done using the computed propensity score $p(X)$. The PSM estimates from this study suggest that knowledge of GI led to higher probability of participating in local organizations and civic activities.

Since the treatment effect estimation is done only using the matched units, the resulting estimate is the ATT. Further assumptions required to conduct PSM are common support and unconfoundedness. Common support ensures that treated units have untreated units "nearby" in the propensity score distribution. Common support can be visualized by plotting histograms of treated and untreated units across the propensity score distribution. The expectation is to see a significant overlap, suggesting a "good match" as shown in Fig. 2.7. The unconfoundedness assumption implies that program participation is determined solely by observed characteristics. This is a strong assumption, and a limitation is that there is no statistical test to prove that there are no unobserved characteristics that affect program participation. However, there are ways to conduct sensitivity analysis to unobserved confounders.

An evaluation of forest protection policies illustrates the use of PSM in assessing environmental policies that suffer from selection bias. Nelson and Chomitz (2011) addressed the fact that protected areas are more concentrated on lands that are unattractive to agriculture, which typically are remote areas with higher slopes and higher elevations because it is easier for governments to implement protection where population density is low and there is less objection (Fig. 2.8).

In such a scenario, an unbiased comparison of deforestation rates between protected and unprotected areas would overestimate the effects

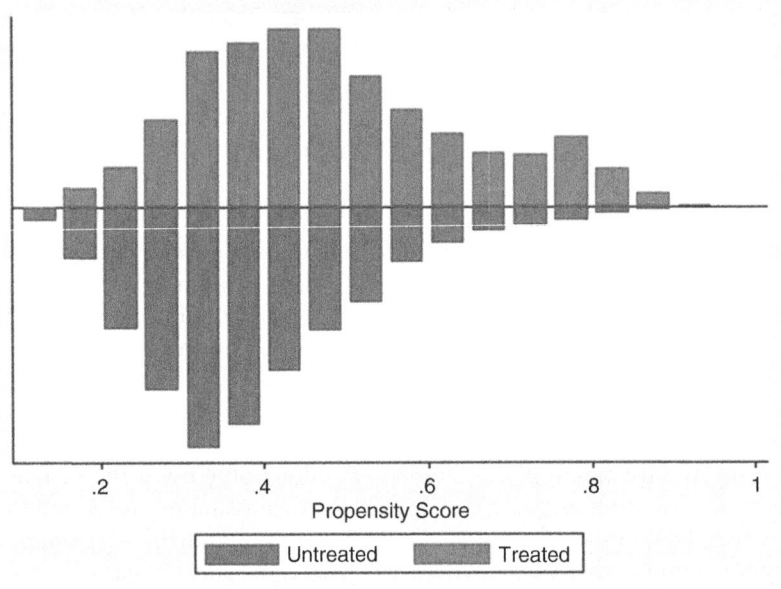

Fig. 2.7 Frequency distribution of treated and untreated units on common support. (Source: Capuno and Garcia 2010)

Fig. 2.8 Protected areas established by 2000. Protected area category: strict (green), multiple use (yellow), indigenous (pink). (Source: Nelson and Chomitz 2011)

of protection. The study used data from developing countries in Latin America, Africa, and Asia and constructed a counterfactual by matching the protected areas and unprotected areas using matching criteria of distance to road network, distance to major cities, elevation and slope, and rainfall. The results showed that the incidence of deforestation is much less in the protected areas than in the unprotected areas.

The study compared the effects of forest protection policy in strictly protected areas, which allow only conservation-related use; multiple-use protected areas, which allow some sustainable use by local inhabitants; and indigenous areas. The general finding was that forest protection policy in multiple-use protected areas was at least as effective as strictly protected areas, suggesting that global environmental goals and local productive activities are compatible. The policy implication derived from this valuable evidence is that setting policies such that there are variations in land use restrictions can be effective in biodiversity conservation and climate change mitigation.

Advantages and Limitations of PSM
PSM is a useful tool to estimate program impact in that it can be applied retrospectively as long as the appropriate data are available. It is desirable to have baseline data, but matching can still be conducted with only post-program cross-sectional data. When there is no baseline data, however, finding all relevant observed covariates is typically a challenge. Further, satisfaction of the common support assumption requires having a large number of treated and untreated units so that a substantial region of common support can be found (usually a large data set). Moreover, as discussed above, unconfoundedness is a strong assumption to make, and

therefore conducting sensitivity analysis to bias from unobserved factors becomes necessary.

Choosing an Impact-Evaluation Method

We have reviewed a number of methods, each of which comes with its own advantages and limitations. How does an evaluator choose which method is best suited to evaluate a particular program? Important questions need to be asked to help determine the most suitable method.

(i) What are the available resources and constraints?
Randomized experiments, by their very nature, are resource and time intensive. Resources needed include financial support and trained man power. A well-designed experiment in a resource-poor environment is bound to fail. Experiments also require pre-intervention or baseline data and a series of post-intervention surveys to be able to capture the treatment effects. While quasi-experiments are less demanding on time and financial support, they still require trained man power to conduct careful econometric analyses. Further, quasi-experiments require good-quality primary or secondary data that are either cross-sectional or panel and have a large sample size, so that the estimates have internal and external validity. Adequate planning and resources are necessary to collect large-scale, nationally representative surveys or panel data.

(ii) Who are eligible units and how are they selected?
Especially in the case of choosing a quasi-experimental method, it is important to know whether there is a well-defined eligibility rule and whether the eligible and non-eligible units complied with the rule.

(iii) What is the nature and stage of the program being evaluated?
In choosing a suitable evaluation method, knowing the scale of the program is helpful. If it is a pilot program or a small-scale intervention, then conducting a randomized experiment might be feasible. In the case of a program that will be nationally rolled out, it may not be feasible to randomize. There are some examples of conducting RCTs at scale, but these require buy-in from policy-makers at the highest level and significant resources (Muralidharan and Niehaus 2017). Therefore, quasi-experimental methods might be

more suitable if appropriate planning and study design is done to collect baseline data. Yet another consideration is the implementation stage of the program. If the program has not commenced, then it might be possible to randomize and collect baseline data. However, if the evaluation is being done ex post, which is mostly the case, then only quasi-experimental methods are suitable.

(iv) What are the outcomes of interest?

A standard way of thinking about outcomes or indicators is that they have to be SMART—specific, measurable, attributable, realistic, and time bound. If the outcomes are not specific or relevant to the objectives of the program, then evaluating them may not be appropriate at all. For quantitative or econometric IEs, it is also necessary that the outcomes are measurable or operationalizable. Further, changes in the outcome need to be attributable to the program to justify conducting an IE. This again emphasizes the relevance of the indicators. Outcomes also need to be realistic in that they are actually achievable through program implementation. In addition, they have to be time bound, that is, evaluators and policy-makers should know when to expect the program to result in the expected outcomes. This may determine whether a quasi-experiment using cross-sectional data is sufficient or whether long-term follow-up, either through experiments through or panel data, is required.

Choosing an appropriate evaluation method by no means necessitates that only one method be used. In fact, combining methods might be a good way to increase the statistical validity of the estimated treatment effects. It is almost a norm to use IV in experiments where compliance is imperfect or where program take-up is driven by self-selection. More and more evaluations are combining methods such as DID and IV or PSM and DID to increase the internal validity and robustness of their estimates. In doing so, it is also important to examine whether the program implementation satisfies the assumptions and conditions of the chosen methods. Table 2.2 summarizes key features of the methods we discussed in this chapter.

Table 2.2 Comparison of key features of empirical evaluation methods

Methodology	Description	Key assumption	Data requirements	Advantages	Limitations	Parameter(s) of interest
Randomized controlled trial (RCT)	Eligible units are randomly assigned to a treatment or control group using an experimental study design	Treatment and control groups are statistically identical in expectation with respect to observed and unobserved characteristics	– Data on treatment variable – Post-treatment outcome data for treatment and control groups – Pre-treatment data on outcome and other characteristics for treatment and control groups to check balance	Generates internally valid impact estimates under the weakest assumptions	– Compliance influences the validity of randomization – Randomization is not always politically feasible – Requires prospective planning and design	Intent to treat (ITT), average treatment effect (ATE), average treatment effect on the treated (ATT), local average treatment effect (LATE)
Instrumental variable (IV)	An IV is used to generate exogenous variation in the endogenous treatment variable	The instrument affects the treatment variable but does not directly affect outcomes (exclusion restriction)	– Data on treatment variable – Data on IV – Post-treatment data on outcome and other characteristics for all units	– Can be applied retrospectively – Estimated causal effects are unbiased even in the presence of imperfect compliance	IV exogeneity cannot be statistically tested except under specific condition of multiple IVs	Local average treatment effect (LATE)

(continued)

Table 2.2 (continued)

Methodology	Description	Key assumption	Data requirements	Advantages	Limitations	Parameter(s) of interest
Difference-in-differences (DID)	Estimates the change in outcome over time between treatment and control groups	In the absence of the treatment, changes in outcome for the treatment and control groups are not different (parallel trends)	– Data on treatment variable – Pre- and post-treatment data on outcome and other characteristics for both treatment and control groups	– Controls for entity- and time-invariant unobserved characteristics – Can be used with repeated cross-sections or panel data	Entity- and time-variant unobserved characteristics cannot be controlled for	Average treatment effect (ATE)
Regression discontinuity design (RDD)	Eligible units are determined by a cutoff based on a continuous (running) variable on which the population can be ranked and which is systematically related to the assignment of the treatment	Units that are immediately above the cutoff and immediately below the cutoff are statistically identical	– Running variable and eligibility cutoff to determine treatment status – Post-treatment outcome data and other characteristics for all units	– Does not require any eligible units to be untreated for the purposes of the impact evaluation – Eligibility cutoff can be either sharp or fuzzy	– Treatment effects around the discontinuity may not be generalizable to the entire treatment group – Estimates can be sensitive to inclusion of different functional forms	Local average treatment effect (LATE)

(continued)

Table 2.2 (continued)

Methodology	Description	Key assumption	Data requirements	Advantages	Limitations	Parameter(s) of interest
Propensity score matching (PSM)	Creates control group from nonparticipants based on matched observed characteristics of the treatment group	There are no characteristics other than the ones used for matching that affect the treatment status	– Data on treatment variable – Post-treatment outcome data and other characteristics for all units	Can be applied retrospectively	Requires large sample size for valid matching	Average treatment effect on the treated (ATT)

Source: Authors' illustration

Note: Treatment variable refers to policy or program variable whose effect the researcher wishes to evaluate

Challenges in Conducting Impact Evaluations

IEs of programs and policies can be valuable inputs into the assessment of how goals of sustainable development are being planned for and met. Aside from challenges of scope, formulation, and presentation of the key issues, technical, organizational, and political challenges can seriously impede the IE process.

Technical Challenges

Technical capacity includes experts who have skills in data collection, data management, and data analysis. In most developing and less-developed countries, training in social sciences, public policy, and quantitative skills is still lacking. Governments and organizations in these countries often have to rely on aid agencies or external evaluators, who may lack local knowledge. Consequently, methods and indicators used for evaluation may not be suitable to the country context, and the evaluation results may not be useful for decision-making purposes.

Policy-makers might support IEs to gain political credibility, but without trained manpower, this may not be feasible. Overcoming these technical challenges requires building relevant human capital and skills.

Organizational Challenges

Organizational capacity refers to administrative coordination as well as financial resources. IEs are rarely institutionalized, in that they do not follow a systematic approach in identifying, implementing, and using evaluations to inform policy decisions (Bamberger 2009). This requires buy-in and participation from all levels within the organization. This remains a challenge as officials may not view participating in evaluations as part of their responsibilities, especially if tenure and promotion are not linked with achieving program outcomes. Conducting relevant, high-quality, and timely evaluations requires close coordination and alignment of goals among policy-makers, organizations, and the evaluators or technical experts.

A further organizational challenge is budget or financial resources. Integrating IE ex ante in policy design requires committing a significant amount of resources to conduct consultations with various stakeholders, collecting pre- and post-policy data, and conducting and disseminating findings. While this is the ideal case scenario in IEs, resource challenges mean that evaluation is usually done ex post.

Political Challenges

While technical and organizational challenges can be addressed by investing in training and organizational learning, overcoming political constraints can be particularly difficult. Policy outcomes can have significant implications on voter preferences and aid agency assessment. Policymakers may therefore be reluctant to conduct IEs, cherry-pick areas where an evaluation can be conducted, or refuse to accept findings from rigorously and independently conducted evaluations because they do not align with voter expectations.

Organizations may have great interest in assessing whether they have achieved their intended objectives. However, if they directly conflict with political interests, then policies may never be put under the evaluation scanner. These challenges defeat the very purpose of conducting IEs. In extreme situations they can make it impossible to conduct any evaluations.

Conclusions

Evidence-based policy-making calls for the use of findings from IEs, whose scope can vary a great deal depending on the questions asked and the availability of data and other resources. The key value added by IE is in delineating how much of an impact can be attributed or causally linked to a specific policy. While applying IE to understand the effectiveness of policies pertaining to inequality, environmental protection, and governance is thought to be challenging, we demonstrate through real-life policy examples how these tools can be applied to address these big issues.

Often, evaluators are at variance when it comes to "attribution" versus "contribution." IE places clear emphasis on causal attribution. However, when an intervention is complex and involves multiple stakeholders and various aspects, such as economic tools, institutional changes, and social reforms, it might become challenging to attribute changes in outcome to one stakeholder or one aspect alone. At most, evaluators can identify various factors that contribute to the overall outcome.

Contribution analysis can be conducted using logical frameworks and qualitative methods such as in-depth case studies and participatory assessment involving different stakeholders, and it can help to understand what value is added by specific stakeholders or individual components of an intervention to the overall outcome. However, contribution and attribution need not be conflicting objectives of the evaluation exercise. In fact, contribution analysis can potentially form the basis of future IEs and thus be more complementary to attribution analysis.

It might be farfetched to suggest that one experiment or quasi-experiment can provide all the answers to complex problems that lie at the core of sustainable development. However, cumulative knowledge accumulated through multiple evaluations conducted in multiple contexts will enable policy-makers to provide answers that are rigorously grounded in evidence.

Notes

1. Expectation or expected value refers to the mean of a random variable.
2. See Hahn, Todd, and Van der Klaauw (2001) for details.
3. See Angrist, Imbens, and Rubin (1996) for detailed discussion on the methodology.

Bibliography

Angrist, Joshua D., Guido W. Imbens, and Donald B. Rubin. 1996. "Identification of Causal Effects Using Instrumental Variables." *Journal of the American Statistical Association* 91 (434):444–455.

Bamberger, Michael. 2009. Institutionalizing Impact Evaluation Within the Framework of a Monitoring and Evaluation System. Washington, DC: World Bank.

Cameron, Drew B., Anjini Mishra, and Annette N. Brown. 2016. "The Growth of Impact Evaluation for International Development: How Much Have We Learned?" *Journal of Development Effectiveness* 8 (1):1–21.

Capuno, Joseph J., and M. M. Garcia. 2010. "Can Information about Local Government Performance Induce Civic Participation? Evidence from the Philippines." *Journal of Development Studies* 46 (4):624–643.

Chen, Yvonne, Namrata Chindarkar, and Yun Xiao. 2019. "Effect of Reliable Electricity on Health Facilities, Health Information, and Child and Maternal Health Services Utilization: Evidence from Rural Gujarat, India." *Journal of Health, Population and Nutrition* 38 (7):1–16.

Chen, Yuyu, Avraham Ebenstein, Michael Greenstone, and Hongbin Li. 2013. "Evidence on the Impact of Sustained Exposure to Air Pollution on Life Expectancy from China's Huai River Policy." *Proceedings of the National Academy of Sciences* 110 (32):12936–12941.

Chetty, Raj, Nathaniel Hendren, and Lawrence F. Katz. 2016. "The Effects of Exposure to Better Neighborhoods on Children: New Evidence from the Moving to Opportunity Experiment." *American Economic Review* 106 (4):855–902.

Deaton, Angus, and Nancy Cartwright. 2016. "Understanding and Misunderstanding Randomized Controlled Trials." *Social Science & Medicine* 210:2–21.

Dinkelman, Taryn. 2011. "The Effects of Rural Electrification on Employment: New Evidence from South Africa." *American Economic Review* 101 (7):3078–3108. doi: https://doi.org/10.1257/aer.101.7.3078.

Duflo, Esther, Rachel Glennerster, and Michael Kremer. 2007. "Using Randomization in Development Economics Research: A Toolkit." In *Handbook of Development Economics*, 3895–3962. Amsterdam: North Holland.

Hahn, Jinyong, Petra Todd, and Wilbert Van der Klaauw. 2001. "Identification and Estimation of Treatment Effects with a Regression-Discontinuity Design." *Econometrica* 69 (1):201–209.

Mauro, Paolo. 1995. "Corruption and growth." *Quarterly journal of economics* 110 (3):681–712.

Meyer, Breed D. 1995. "Natural and Quasi-Experiments in Economics." *Journal of Business & Economic Statistics* 13 (2):151–161.

Muralidharan, Karthik, and Paul Niehaus. 2017. "Experimentation at Scale." *Journal of Economic Perspectives* 31 (4):103–124.

Nelson, Andrew, and Kenneth M. Chomitz. 2011. "Effectiveness of Strict vs. Multiple Use Protected Areas in Reducing Tropical Forest Fires: A Global Analysis using Matching Methods." *PloS ONE* 6 (8):e22722.

Ravallion, Martin. 2018. Can high-inequality developing countries escape absolute poverty? *Center for Global Development Working Paper no. 492*. Washington, DC: Centre for Global Development.

Slough, Tara, and Johannes Urpelainen. 2018. Public Policy Under Limited State Capacity: Evidence from Deforestation Control in the Brazilian Amazon. *Mimeo*.

United Nations. 2016. Global Sustainable Development Report 2016. Chapter 2 "The infrastructure – inequality – resilience nexus". New York: Department of Economic and Social Affairs, United Nations.

White, Howard, and David A. Raitzer. 2017. Impact Evaluation of Development Interventions: A Practical Guide. Manila: ADB.

WHO (World Health Organization). 2014. Access to Modern Energy Services for Health Facilities in Resource-Constrained Settings: A Review of Status, Significance, Challenges and Measurement. Geneva: World Health Organization.

Open Access This chapter is licensed under the terms of the Creative Commons Attribution 4.0 International License (http://creativecommons.org/licenses/by/4.0/), which permits use, sharing, adaptation, distribution and reproduction in any medium or format, as long as you give appropriate credit to the original author(s) and the source, provide a link to the Creative Commons licence and indicate if changes were made.

The images or other third party material in this chapter are included in the chapter's Creative Commons licence, unless indicated otherwise in a credit line to the material. If material is not included in the chapter's Creative Commons licence and your intended use is not permitted by statutory regulation or exceeds the permitted use, you will need to obtain permission directly from the copyright holder.

CHAPTER 3

The Picture from Cost-Benefit Analysis

Abstract This chapter focuses on one of the oldest techniques of economic evaluation, cost-benefit analysis (CBA). It takes the readers through the steps involved in conducting a CBA. In addition to conventional steps to be followed, it underscores the use of the technique in examining inclusion and environmental sustainability. It also includes case studies highlighting the application of CBA in influencing policies aimed at achieving sustainable development.

Keywords Externalities • Equity • Distributional weights • Net present value • Discount rate • Net social benefit

Decision-makers faced with competing alternatives often need to answer the question whether it is worth investing taxpayer or aid dollars in the pursuit of projects aimed at sustainable development. This is particularly so when longer-term goals seem to come at the expense of short-term economic or political objectives. Choosing to provide rural households with 24-hour electricity may seem like an obvious policy choice keeping in mind broader Sustainable Development Goals. However, when providing household electricity may come at the expense of meeting other priorities, is it still the right decision?

Making decisions, especially involving trade-offs, requires an exposition of the benefits and costs of a project, including the identification of the main cost and benefit components and their valuation in monetary terms. Cost-benefit analysis (CBA) enables decision-makers to weigh the benefits and costs that accrue to the society, beyond individual entities, in comparable monetary units. In doing so, it helps them to allocate resources in the most efficient manner.

Why Use CBA?

CBA is one of the most prominent and widely used evaluation and decision-making tools in public policy. Impact evaluation (IE), as discussed in the previous chapter, focuses on the contribution that can be attributed to a program, and it can be used to identify causality by comparing the outcomes of those benefiting from the project with the counterfactual. CBA has distinctive features that make it complementary to IE.

First, CBA is prospective, in addition to being useful for looking at results retrospectively. It can be used to make projections and calculate the net benefit in terms of present value. This information can allow policymakers to not only assess whether a project provides enough net benefits to warrant investing limited resources in it, it also provides a measuring stick to help them choose among alternative uses of resources. It therefore provides a firmer basis for the choices made.

Second, CBA assigns values to the benefits and costs of projects. While IE is intended to measure how much the treated individuals are better off (or worse off) compared to the case where there is no intervention, it does not directly consider how much costs were incurred to implement a project or how much have the beneficiaries benefited in monetary terms.

Third, the analysis, in principle, covers the range of benefits and costs, whether they have market prices or not. Many projects generate intangible benefits, which may be difficult to monetize. CBA includes techniques to value such unpriced benefits, both current and future, in present-dollar terms. On many of the SDGs, placing a value on intangible, indirect, and unintended attributes could be crucial. For instance, related to SDG 3 (good health) is an assessment of indirect health benefits of rural electrification such as improved health systems (Chen et al. 2018). However, an intangible outcome of improved rural water supply that is related to SDG 6 (clean water and sanitation) might be increased subjective well-being (Mahasuweerachai and Pangjai 2018). Some recent CBA studies rigorously deal with indirect costs and benefits of projects with regard to environmental protection, another important pillar of SDGs (IED 2016; Rojas-Bacho et al. 2013).

These features have been widely appreciated in economics, particularly in sectors which require ex-ante assessments of large-scale investments, such as energy, transportation, and urban/rural development projects. The use of CBA in some institutions, however, has witnessed a decline over the past few decades. A study by the Independent Evaluation Group (2010b) finds that the percentage of projects for which a CBA was performed (using an economic rate-of-return estimate) declined from 70 percent in the early 1970s to about 30 percent in early 2000s. This was partly explained by a relative shift from the sectors like energy and transport that usually apply CBA to those like education and health that conventionally are hesitant to do so.

Conducting CBA requires data to estimate the likely benefits and costs. A constraint may be the lack of readily usable data on the benefits of certain interventions, for example, health gains from reducing water pollution or improving sanitation. It would help to invest in data collection throughout the planning and implementation stages of projects. Important would be efforts to strengthen the capacity of governments and organizations in this respect as many also lack proper record-keeping processes, making it difficult to use data from previous projects.

It may be more difficult to quantify benefits and costs in some sectors. In recent decades, both governments and aid agencies have been increasingly investing in social sectors such as education and health that may face relatively more data challenges. In some instances, the argument for not utilizing CBA also points to the difficulties in quantifying intangible or non-monetary benefits such as empowerment or improved life satisfaction.

However, it would pay to expand the use of CBA across sectors, particularly in light of advances in data and estimation. Increased sophistication in conducting IEs has meant that it is now possible to get estimates of effects of projects that have already been implemented. These estimates can feed into CBA of future projects and help overcome the oft-cited data limitations, including intangible benefits. This also opens up the field for applying CBA across themes from social inclusion to environmental protection to governance.

STEPS IN CONDUCTING CBA

Among the steps involved in conducting a CBA, we need to think most carefully about the costs and benefits of the intervention. By eventually aggregating the associated costs and benefits, CBA can guide investing in a project that has the prospect for enhancing sustainable development. In discussing the costs and benefits, let us consider a rural electrification intervention using the case of an actual project implemented in the district of Ribáuè, Mozambique (Mulder and Tembe 2008).

In designing any intervention, it is essential to identify the policy problem first. Access to modern energy services in Mozambique was still very low, with the vast majority of the population relying entirely on traditional biomass to meet their energy needs at the time of the writing about the project (see also World Bank 2018 for comparative figures). The large gap between urban and rural areas was a pressing concern. Mozambique at that time also exported about as much electricity as it imported. Figure 3.1 shows the geographical setting.

In these circumstances, the payoffs to investing in increasing the coverage of electricity are likely to be high. At the same time, in view of differing costs affected by various factors, it is important to consider alternatives carefully.

The government of Mozambique adopted a National Master Plan for electrification in 2004 in which one of the targets was raising the electricity access rate to 20 percent by 2020. The total investment amounted to US$850 million, of which some US$260 million was allocated toward transmission projects and some US$475 million was allocated toward distribution projects including rural electrification projects, like the one being considered here.

Policy alternatives need to be defined in the context of the problem for which the investment is being sought. The set of alternatives would include the baseline case and other alternatives that are expected to help solve the problem. The baseline case would be the counterfactual assuming there are no changes in policies. Each alternative may vary in inputs used, target area or population covered, or implementation timing. The alternative description includes actions, resources required, and expected results. The alternatives in the case of increasing electricity access in rural areas might be to invest in microgrids or solar energy rather than expanding national-grid coverage.

Should resource constraints limit the coverage of policies or projects, specific populations or regions may be given priority over others. An observation in the case of expanding rural electricity access has been that poor households are less likely to benefit from the intervention, as they may not be able to pay even the minimum user charges. This essentially excludes them as beneficiaries. Alternatively, the design of the program might be such that poor households are provided financing and increasing

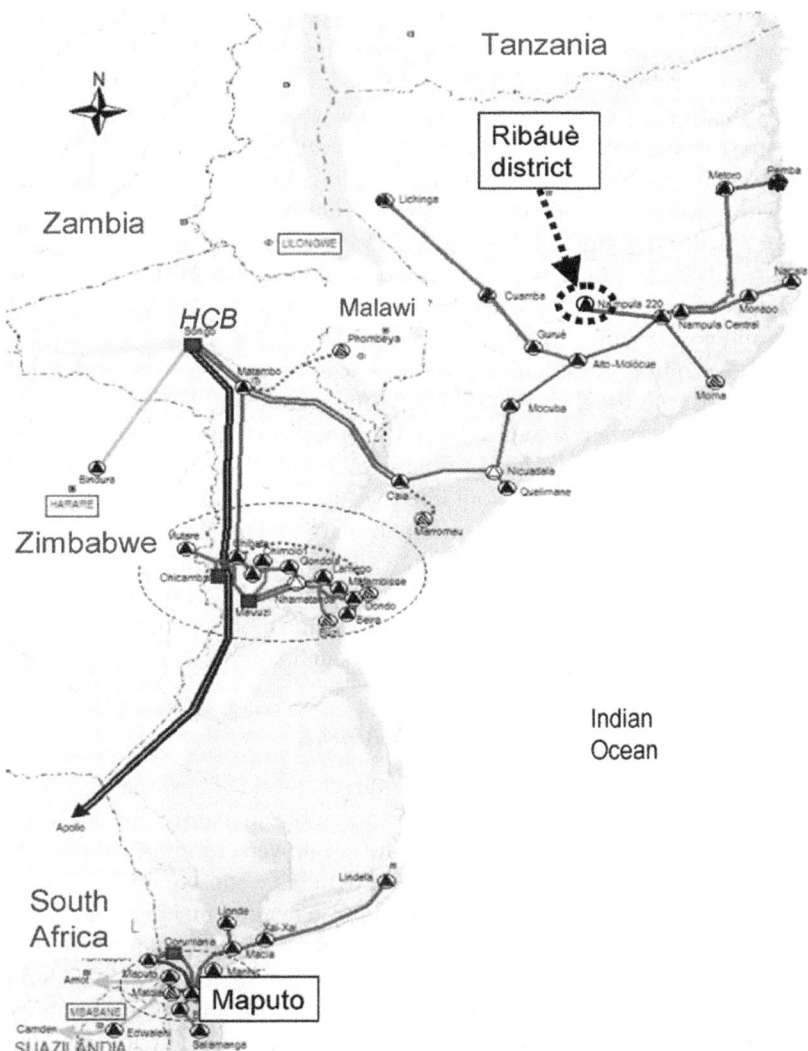

Fig. 3.1 Mozambique electricity system and Ribáuè district. (Source: Mulder and Tembe 2008)

block tariffs are applied for non-poor households based on usage. In this case, benefits accruing to both the poor and non-poor households will have to be accounted for.

Estimating costs and benefits is not always simple. Costs that will be incurred in implementing the rural electrification program may be valuated relative to costs that would have been incurred even if the program had not been implemented. Expected benefits to program households would include tangible and intangible benefits. Spillover benefits to households outside the program areas could be important. All these must be considered to ensure that the identified impacts are the incremental benefits and costs relative to the counterfactual.

An overriding consideration involves whose viewpoint or "standing" the analysis tries to represent (Whittington and MacRae 1986). Persons or entities may be given standing by having their preferences or viewpoints counted as the basis for decision-making that aims to maximize welfare. An environmental example would be the case of a pollution control project that abates carbon emissions where it matters if net benefits are sought to be maximized from the point of view of the people living in a locality or a region or the world. Problems of standing also arise in the valuation of life, the consideration of future generations and nonhuman entities, and distributional concerns and weighting of benefits.

Identifying Benefits

CBA needs to incorporate direct and indirect social benefits that accrue to society whether they are tangible or intangible in nature. In addition, externalities, which are positive or negative spillovers, can be an important part of CBA as will be discussed in more detail in a subsequent section.

Here, we identify potential social benefits generated by the rural electrification project in Mozambique. Investing in electrification is directly related to SDG 7 on affordable and clean energy, as well as indirectly to SDG 8 on decent work and economic growth and to SDG 9 on industry, innovation, and infrastructure. The most direct benefit of domestic electricity would be lighting. Benefits from switching to electricity can be calculated by comparing costs of using alternative sources of lighting such as kerosene.

Use of electricity might save time spent on household chores. The time saved can be used either on productive activities, such as wage work, or leisure. This can be valued at the opportunity cost of wage work using the prevailing average market wage rate.

Improved domestic electricity can potentially lead to an increase in household enterprises such as running a small shop or a sewing business. The value of such benefits is the incremental revenue from the enterprise. Households might save on energy costs by substituting electricity for kerosene to meet their cooking and lighting needs. These savings would need to be incorporated in computing net household revenue.

Indirect benefits of rural electrification might include improved health. This might be due to operational efficiency of health-care facilities and longer hours of operation of the clinics. Health benefits might also accrue from reduced indoor air pollution as households shift away from using kerosene and fuelwood for lighting and cooking and switch to electric cook stoves and lightbulbs. Environmental and climate benefits are increasingly noted in the case of energy reforms and energy projects that aim for energy efficiency or a switch to cleaner fuels (see, e.g., IEG 2010a for an application of CBA to energy efficiency projects in China).

Electrification may also reduce fertility as households have alternate sources of recreation or receive family-planning information through television. Additional indirect benefits may include education as children can study and do homework after dark or schools are able to invest in better learning technologies. These benefits can be valued using out-of-pocket health-cost savings, statistical value of a life-year, costs of implementing family-planning programs, and increase in potential wages after school completion.

Much of the additional electricity production in Mozambique is aimed at increasing exports. At the time of the study, in Ribáuè district, a large volume of electricity was consumed by mills producing and exporting cotton fabric and maize. Efficiency of cotton processing and maize milling might improve with electricity, which in turn might increase production levels as well as incomes. Local cotton fabric and maize mills might also make additional savings in energy costs by substituting electricity for diesel.

A side effect might be increased tax revenue from increased household enterprises and commercial activity, which is a transfer from households and businesses to the government. Spillover effects might include environmental benefits or costs that must be counted. Electrification increases energy consumption, which in turn may increase emissions. However, emissions can be partially offset by a shift toward cleaner fuels such as using electricity for lighting instead of kerosene. Other effects might include a reduction in rural-urban migration due to increased economic activity and more job opportunities in rural areas. This in turn might ease the congestion in urban areas.

Identifying Costs

As with benefits, identifying the costs associated with a project is not always straightforward. The total cost for each alternative is the increase in the cost relative to the counterfactual, which includes investment and recurrent costs. Investment costs include the necessary costs to implement the project, while recurrent costs include costs in areas such as operation and maintenance.

For expanding rural electrification, the investment costs include those for land, physical infrastructure, technology, and manpower required to expand the national grid. Recurrent costs include those for manpower and public works necessary to maintain the technology and infrastructure. Not included would be sunk costs, which were already incurred or would have been incurred regardless of project implementation and could not be reversed. There is no opportunity cost associated with sunk cost. These costs do not affect the decision of whether to implement the project, or which alternative to select, and are therefore not included in the CBA.

If the government was investing in developing new technology to reduce transmission and distribution losses from the grid at a national level, and if the same technology were to be used to expand rural electricity access, then the R&D cost is essentially a sunk cost and is not be included. However, if further R&D were required specifically for the rural electrification project, then it would be included as an investment cost.

Identification and valuation of physical inputs might be straightforward. What is difficult is determining how much of the inputs will be needed and when or for how long they are used. Once these quantities are known, they are valued at their opportunity cost, which is the difference in return between the current use of inputs and the return if the inputs were put to their next best use. In cases where market prices do not reflect opportunity costs due to market failures, shadow prices may be used. A shadow price is a proxy value of the good, usually inferred from stated preferences such as willingness-to-pay (WTP) or willingness-to-accept.

Identification of labor inputs is relatively easy. Valuation generally uses prevailing market wages. Shadow wages may be used in the case of market inefficiency. These are valued using the forgone output if labor (assumed to be fixed in supply) is taken away from other sectors or projects and used in the construction of electricity infrastructure.

Cost estimation is generally based on the assumption that there will be no modifications to the project design. A contingency allowance allows for any distortions in design or implementation schedule, which may add to the base cost estimate. Contingencies are generally of two types: physical contingencies and price contingencies. Physical contingencies are to cover for physical uncertainties such as increase in the use of real goods and services. They are often calculated as percentages of base costs.

Price contingencies are to cover for inflation and price uncertainties. If, say, owing to topography, the material requirements for constructing high-voltage transmission towers increase, the cost difference would be covered by physical contingencies. On the other hand, if the price of steel required for construction of transmission towers increases globally, then these changes would be covered by price contingencies.

A project may generate indirect costs. For instance, electrification of households may negatively affect agricultural production if it comes at the cost of rationing electricity supply to farms. The loss in crop production can be valued at market prices, assuming a competitive market exists. Other indirect costs may include the effect of television viewing on children's propensity to read or study (World Bank 2000). This can be valued using lost future wages.

Valuation Techniques

Placing values on the principal benefits and costs is a challenge. The valuation technique needs to be selected carefully for each identified cost and benefit so that true social value is derived. Valuation is less complicated when competitive markets exist for the goods and services being included. There are ways and means to value benefit and costs, for instance, improvements in transportation that save time, which people are willing to pay for through the transportation choices they make.

People increase their consumption of something till the additional benefit equals the market price. In principle, the demand schedule for the product in question provides information on the additional benefit reflecting the monetary value of an increase in consumption. This approach can work for goods and services traded in the market. A good part of CBA can rely on the application of market demand for which historic data might be available.

Other indirect valuation techniques are required under conditions of imperfect or missing markets, as discussed in the examples of benefits and costs in the preceding section (see, e.g., Tolley and Fabian 1998). Valuation techniques commonly used in determining shadow prices include: contingent valuation, hedonic pricing, and travel-cost technique.

We now know that rural electrification brings many unpriced benefits: improved health from better indoor air quality and better education outcomes owing to lighting. Such unpriced benefits can be valued by asking individuals the maximum amount they are willing to pay for the benefits. Jeuland et al. (2015) provide an interesting example of preferences for biomass burning compared to improved cook stoves among rural households in north India, one that also captures the accounting of negative externalities (discussed later). This technique is called contingent valuation. It is a flexible tool in that it can be used to value almost anything, although it works best to value benefits from goods and services that individuals can identify and understand.

A major challenge of this technique is bias, which may occur because it depends on the responses given to survey questions. One source of bias is the level of the respondents' knowledge about the goods or services in question. Bias may also occur if the individuals are not familiar with placing monetary values on such things as environmental goods and services, in which case the stated WTP may not reflect the true value.

Taking an example from concerns over indoor air quality, the simplest application of contingent valuation would be to ask respondents what is the maximum price they are willing to pay to mitigate indoor air pollution. More sophisticated ways include presenting respondents with a range of values or double-bounded dichotomous choices and testing their sensitivity to different price levels. Once data are collected, the average WTP and the conditional demand curve at each price level can be traced out.

Hedonic pricing uses the market prices of goods and services under the assumption that these prices reflect the value that people place on their characteristics. It is often used to value environmental goods such as air, water, and land quality. For instance, if two houses, house A and house B, are identical except for the air quality in their respective neighborhoods, the price difference between the two houses can be considered as the price the consumer is willing to pay for good air quality (see Tan Soo 2017 for an example). Hedonic pricing can also be used to infer health benefits or mortality risks by taking the difference in market wages of jobs in green sectors versus those in polluting sectors.

The strength of this method is that it uses the actual market prices and characteristics of the goods and services. The limitation is that it assumes that people are aware of the less tangible attributes of a good, such as the environment and its health attributes or consequences. Hedonic pricing requires that changes in attributes be linked to WTP; and its use requires data availability and a high degree of technical expertise.

In the valuation of housing or land in electrified villages, the per-square-foot price in villages with 24-hour supply may be compared to the price in villages with no or intermittent power supply. The difference in prices reflects WTP for improved access to electricity.

Mostly used to value the benefits of recreational facilities, the travel-cost technique can also be applied to value environmental and health goods. This technique focuses on the access cost and assumes that an individual will access the facility if the benefits gained outweigh the total travel costs (including opportunity cost of time). Consumers' WTP to visit a recreational facility is thus estimated using the number of trips that they make at different travel costs.

We identified one of the indirect benefits of rural electrification as improved health owing to better operational efficiency of health facilities. To value the health benefits, we can utilize the travel cost that individuals actually incur to visit a clinic with better facilities and equipment in the absence of such a facility nearby.

Avoid Double Counting

Double counting is a common error in aggregating benefits and costs. As a rule, a particular benefit or cost should not be counted more than once. If a project produces intermediate goods that are used as inputs to some other downstream products, the total benefit should only be counted once.

In our rural electrification example, better access to electricity often leads to an increase in time available in terms of labor supply as households save time on chores. If the households have already valued their time use when stating their WTP for electricity, including time use benefits as additional benefits would amount to double counting. Similarly, counting increases in household enterprises owing to increased time available in labor supply, and counting time inputs for household enterprises, would be double counting. Obtaining total benefits can therefore be difficult, as chances of either double counting or underestimating are high, especially if most of the benefits are indirect or intangible.

Taxes, subsidies, and government charges could also be a source of double counting if not accounted for properly. Valuation should be based on the concept of WTP and opportunity cost. For example, the private cost of electricity should be valued at what the consumers are willing to pay inclusive of taxes or transfers to the government. However, cost to the society should be valued using the opportunity cost of production, that is, the price of inputs used to produce forgone outputs, which is exclusive of taxes.

Computing Net Social Benefit

The objective function of CBA is to estimate net social benefit. Once we value benefits and costs, we can compute consumer and producer surplus to arrive at net social benefit (Harberger 1971). Continuing with our rural electrification case study, let us consider two sources of lighting available to households: electricity and the existing source, usually a kerosene lamp.

The valuation of benefits is often done using WTP. Figure 3.2 shows a demand curve derived by the price of lumens (total quantity of visible light emitted by a source); the quantity of kerosene consumed, Q_k at price P_k; and the quantity of electricity consumed, Q_e and price P_e. The area under

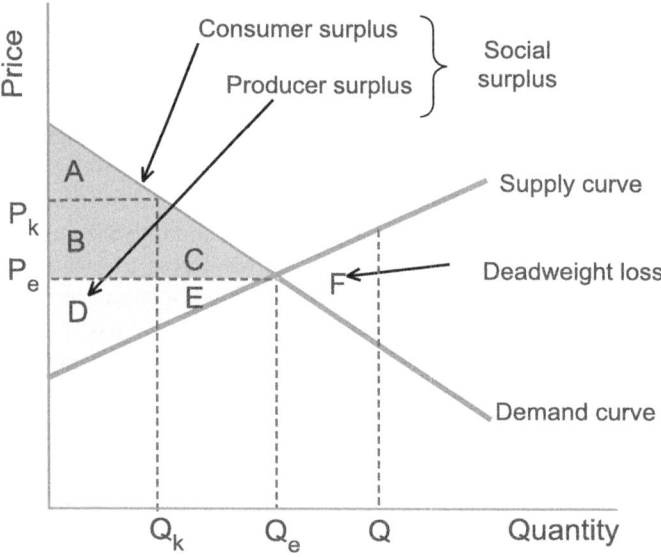

Fig. 3.2 Consumer surplus and producer surplus for kerosene and electricity. Note: P_e is the price of electricity from the grid; P_k is the price of kerosene; Q_e is the quantity of electricity used from the grid; and Q_k is the quantity of kerosene consumed. (Source: Authors' illustration)

the demand curve is the amount the consumer is willing to pay (see also IEG 2008).

The difference between what one is willing to pay and what one actually pays is the consumer surplus. With the assumptions in Fig. 3.2, the consumer surplus for kerosene would be $(A+B+D)$ minus $(B+D)$, which is A. In the same manner, the consumer surplus for electricity would be $(A+B+C)$. Providing electricity, therefore, increases consumer surplus from A to $(A+B+C)$, producing the additional benefit of $(B+C)$ to consumers.

The cost side reflects the opportunity cost, or what is given up, to produce the good in question. The marginal cost curve shows the additional cost incurred to produce an additional unit of a good, and when summed up for the individual producers, it produces a market supply curve. The area under this curve indicates the total variable cost. What the producers actually receive is the price multiplied by the quantity of the good. And the difference between the total revenue and the total variable cost would be the producer surplus for electricity, which is the triangle area $(D+E)$.

The equilibrium price and quantity determined in competitive markets yields a consumer surplus and a producer surplus, whose sum is the social surplus. This is the net social benefit as it is calculated by subtracting costs (as opportunity costs) from benefits (as WTP). This equilibrium point reflects allocative efficiency because outputs any less than Q_e result in a reduction in social surplus, making at least some people worse off relative to the equilibrium point. For example, the output Q, which deviates from the socially optimum point, will result in less social surplus by the triangle area F, called deadweight loss.

Externalities

Not elaborated thus far, a vital part of net social benefits is estimating social impacts that may arise in the presence of market failures and market imperfections. Referred to as externalities, these spillover effects can be positive or negative. For instance, subsidized electricity provided to farmers may induce them to extract groundwater excessively, thus reducing the amount available for rural households. Less water for domestic consumption can hurt people's health. This negative health externality imposes a cost on society.

Figure 3.3 illustrates how the additional cost of a negative externality shifts the marginal cost to society or the supply curve upward by t_n, caus-

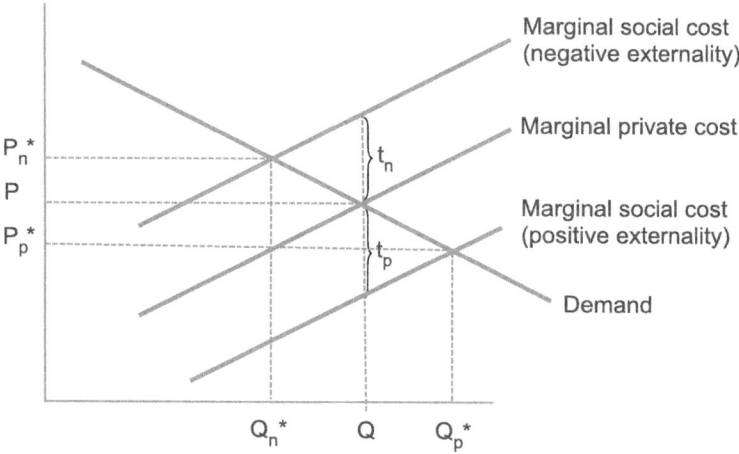

Fig. 3.3 Marginal social cost in the presence of a positive or a negative externality

ing the equilibrium quantity to decline from Q to $Q_n{}^*$ and the price to rise from P to $P_n{}^*$. As a result, net benefits decline. In the same way, a positive externality reduces the costs to society, resulting in downward shift of the marginal social cost curve by t_p, increasing the net social benefits.

Project Benefits and Costs

Identified benefits and costs need to be projected over time. A project-implementation schedule plays a critical role here. Delay in implementation often results in unexpected costs or loss of anticipated benefits. The project may initially bear large benefits that diminish gradually, or it may take some time to fully manifest its effects, resulting in little or no benefit in the first year. For some projects, costs may be large initially but diminish over time, while other projects may require high maintenance costs over the life of the project.

Mulder and Tembe (2008) use data for the rural electrification project over 2000–2005 and project costs and benefits for the period 2005–2020. We can add simulations of the CBA to account for issues such as distributional considerations and uncertainties about the stream of benefits and costs. The study lists the costs and benefits for the project, which are illustrative of estimates that correspond to rural electrification projects more generally. Costs mostly comprise investment costs and operating and maintenance costs for line construction. Direct benefits include saving in energy costs for businesses and households and increase in energy efficiency for cotton, maize, and other enterprises. There are also important indirect benefits in the areas of education and health. For education, benefit estimates were based on measures of the increased number of students finishing school enabled inter alia from offering night classes and better facilities. Estimation of health improvements would have required a valuation of the health gains from increased hours running hospitals and equipment.

Based on the costs and benefits in 2000–2005, the authors provide three scenarios of benefits and costs: "high" in terms of being optimistic, "medium" in terms of being average, and "low" in terms of being pessimistic. On the cost side, these scenarios yielded three sets of projections of the operating costs. On the benefit side, there were three sets of projections of the direct gains in the form of energy savings and processing improvements in cotton, maize, and other mills. The indirect gains in education from the project too had a range of optimistic, average, and pessimistic increases in the number of new students.

The study's estimates show how costs, benefits, and net benefits might evolve up to 2020. Interesting is how the estimates are influenced by the expectation of how direct and indirect benefits evolve, including the gains expected in education. The range of low, medium, and high for costs and benefits provides variations that can inform the projections of cumulative net benefits of the three scenarios. Using the medium estimate, the study finds that the benefits strongly outweigh the costs over the period 2005–2020.

Net Present Value (NPV) of Alternative Scenarios

It is necessary to discount benefits and costs to their present value to account for the time value of resources. Future benefits and costs must be converted into present value terms by applying an appropriate discount rate. Commonly, people tend to discount the value of things whose consumption is in the future. Discounting the future assumes that individuals prefer current consumption to future consumption. The magnitude of discount can also be assumed to increase with time.

Net present value (NPV) is calculated using the discount rate as follows:

$$\text{Net present value} = \sum_{t=0}^{n} \frac{NB_t}{(1+s)^t}$$

where NB is annual net benefit in year t and s is discount rate. As is clear from the equation, NPV is defined as the sum of a flow of annual net benefits, which is converted to a present value.

In appraising public policies whose purpose is to improve benefits to society, CBA uses a social discount rate. This is the social rate of time preference, which is the rate at which society would trade a unit of benefit between the present and the future. Choice of a social discount rate greatly influences the decision-making in selecting one project over another. A high social discount rate means that future benefits and costs count for less; therefore, projects with early benefits are preferred. On the other hand, a low discount rate suggests higher valuation of future benefits and costs. Environmental projects typically adopt low discount rates, as the benefits may manifest in later years.

Some projects, especially those pertaining to environmental protection, have impacts across generations. For their CBA, it may not be enough to apply a constant discount rate over the long time period. Instead, a time-declining discount rate may be considered. The use of a time-invariant discount rate in an evaluation of environmental conservation policy is likely to underestimate the benefits for future generations.

The NPV of a project vitally depends on the discount rate selected. Table 3.1 illustrates how NPV is calculated using different discount rates. For simplicity, annual net benefit is assumed to be a constant US$100 million over ten years. With a constant discount rate of 5 percent, the NPV is US$772.16 million. At 7 percent, the NPV is less than that under the 5 percent discount rate scenario because the future benefits are valued lower.

Next, we apply the time-declining discount rate, which starts at 7 percent and declines over the years to 3 percent. Even for a period as short as ten years, the selected discount rate makes a significant difference in the computation of NPVs. Using the constant discount rates of 3 percent, 5 percent, and 7 percent for an annual benefit of US$100 million, NPVs in 100 years are US$3.16 billion, US$1.99 billion, and US$1.43 billion, respectively. The difference in NPVs is significantly higher when the time period is long. Applying a constant discount rate in analyzing the benefits and costs of projects that are likely to have intergenerational effects might not fully measure the true value of benefits and costs far into the future.

Table 3.1 Calculation of NPVs for alternative scenarios (in million US$)

Year	Annual net benefit $	Present value $		
		Discount rate		
		5%	7%	7%→5%→3%
1	100	95.24	93.46	93.46
2	100	90.70	87.34	87.34
3	100	86.38	81.63	81.63
4	100	82.27	76.29	76.29
5	100	78.35	71.30	78.35
6	100	74.62	66.63	74.62
7	100	71.07	62.27	71.07
8	100	67.68	58.20	78.94
9	100	64.46	54.39	76.64
10	100	61.39	50.83	74.41
Net present values		772.16	702.34	792.75

Source: Authors' illustration

Note: In the last column, it is 7% for years 1–4, 5% for years 5–7, and 3% for years 8–10

The rural electrification project is one such policy intervention that is likely to yield benefits across generations. How individuals value improved health, higher human capital, or cleaner air in a hundred years is unlikely to be reflected in how they currently value the benefits. Selecting a constant discount rate based on the time preference today could lead to rejection of making an investment in rural electrification that potentially has large future benefits.

As the calculation of NPV concerns long-term influences, there are different views on how to incorporate those into the present value. Petri and Thomas (2013) discuss the discount rate and differing views about it in the context of climate change, taking the *Stern Review* (Stern 2006) as a case study. The *Stern Review* uses a normative interest rate of 1.4 percent to calculate the present value of future environmental damages, using the principle that the welfare of all generations should count equally. In contrast, others use a positive rather than a normative rate, say around 6 percent, as observed in market decisions (Nordhaus 2007). This choice has far-reaching consequences.

Table 3.2 presents a comparison of how we can arrive at NPV through market decisions versus normative decisions concerning sustainability. It is

Table 3.2 How prices shape decisions

Price discounting	Market prices	Externality-corrected prices
Market interest rates	Market decisions	Decisions that internalize the external effects of transactions
Normative (low) interest rates	Decisions that internalize the interests of future generations	Decisions that internalize external effects and the interests of future generations

Source: Petri and Thomas (2013)

a matrix of price and discount rate options, with the varying combinations of these two showing how externalities and future effects are valued. The upper-middle cell of the table shows market decisions, that is, transactions based on market prices and interest rates. These do not account for externalities nor for future effects. The upper-right cell shows the effects of price signals that correct for externalities; for example, prices inclusive of carbon taxes. These decisions account for externalities, but from the viewpoint of profitability for today's investors.

The lower cells introduce corrections to interest rates and externalities. In this row, the welfare of future generations is treated similarly to that of the current one. This row would likely require adjustments to lower discount rates from positive to normative levels. Some have argued that committing to such a welfare perspective would require that subsidies be offered to such investments.

The lower-middle cell of the table shows decisions that account for the interests of future generations. The lower-right cell introduces price signals that reflect externalities and accounts for the interests of future generations. That is, it makes decisions sensitive to the interests of others affected by externalities and those living in the future.

Uncertainties and Risks

Clearly, several assumptions go into the identification and valuation of benefits and costs and the computation of NPV. A key question is how net social benefits change if the assumptions, for example on the scope of rural electrification or on market wages, change. We can examine how sensitive predicted net benefits are to changes in assumptions by conducting a sensitivity analysis.

There are different ways to conduct sensitivity analysis that touch on the various parameters involved in CBA. One method is to recalculate the NPV under different sets of assumptions. Based on the best forecast made through the calculations, the range can be set within which some variables could vary. Conducting sensitivity analysis enables policy-makers to make choices under varying levels of risk and uncertainty.

In the case of rural electrification, the NPV can vary according to different estimates of how much time is saved. The benefit from the estimated time saved by using an electrical appliance can be calculated using the prevailing market wage, but there can be variation in the estimate of the market wages used as well. Table 3.3 illustrates how three alternatives of low, best, and high estimates of the time saved and of the market wage rate can produce nine possible alternative NPVs. Both market wages and time saved are crucial factors in that reasonable changes in these variables could significantly alter the NPV computation.

Factoring Social Equity

Even though rising inequality is viewed as a major global challenge, policy-making using CBA for the most part has not explicitly incorporated equity aspects in the analysis. One of the criticisms of CBA is that it focuses on efficiency at the expense of social equity (Kind et al. 2017). This concern arises from the premise in CBA of favoring a project based primarily on aggregate net benefits and the underlying criterion for Pareto improvement.

Concerns about equality may be neglected in the so-called Pareto improvement criterion. Public policies could be intended to make the

Table 3.3 Illustration of NPVs for changes in market wages and time saved

			Time saved by using electrical appliance (hours)		
			Low	*Best*	*High*
			2	3	4
Market wage rate (dollars/hour)	Low	12	NPV1	NPV2	NPW3
	Best	15	NPV4	NPV5	NPV6
	High	18	NPV7	NPV8	NPV9

Source: Adapted from Sinden and Thampapillai (1995)

worse off better, which may not always be a Pareto improvement. Rural electrification is a public policy effort toward greater social inclusion. But there could be situations where such an investment would pass the CBA test only if a dollar increase in the income of a poor rural household is seen to result in a larger increase in social welfare than would a dollar increase in the income of a non-poor rural household.

Most cost-benefit analyses stress the objective of improving social welfare or the well-being of all individuals. But in practice, the welfare implications arising from income differences are not considered. If diminishing marginal utility of money is recognized, then income differences can be accounted for in calculating social welfare benefits. This is especially so for projects with special implications for low- income groups, for example, those dealing with damages from natural calamities that hurt the poor disproportionately.

Considerations of income differences can be realized by assigning distributional weights to various subgroups. Distributional weights reflect the value placed on each dollar received by each group. One general approach is to make the weights inversely proportional to income that would affect the estimated net benefits and favor policies that tend to improve the income distribution.

Applying distributional weights would change the NPVs we previously computed, where equity was not explicitly incorporated. Table 3.4 calculates the NPVs under two alternatives. In both alternative A and alternative B, the annual net benefit is assumed to be US$100 million each year, and the discount rate is 5 percent. This yields an unweighted NPV of US$772.16 million. We now divide the population into poor and non-poor rural households and assume that the benefits received from the electrification project differ for the two groups.

There are several ways to show how results vary according to the weights attached to outcomes affecting particular groups. In one of such ways, consider alternative A, which brings much higher benefits to the non-poor households especially at the beginning of the project, as is often the case with electrification projects. In alternative B, on the other hand, the poor benefit from the project much more than the non-poor. The total benefit is taken to be the same. But from a distributional perspective, the benefits received by the poor might be valued higher than the benefits received by the non-poor, say with a distributional weight of 2 for the poor and 0.5 for the non-poor.

Table 3.4 Calculation of NPVs with distributional weights (in million US$)

Year	Total annual net benefit $	Alternative A					Alternative B			
		Annual net benefit $		Present value $ (discount rate: 5%)			Annual net benefit $		Present value $ (discount rate: 5%)	
		Poor	Non-poor	Unweighted	Weight Poor = 2 Non-poor = 0.5		Poor	Non-poor	Unweighted	Weight Poor = 2 Non-poor = 0.5
1	100	10	90	95.24	61.91		90	10	95.24	176.19
2	100	10	90	90.70	58.96		90	10	90.70	167.80
3	100	10	90	86.38	56.15		90	10	86.38	159.80
4	100	20	80	82.27	65.82		80	20	82.27	139.86
5	100	20	80	78.35	62.68		80	20	78.35	133.20
6	100	20	80	74.62	59.70		80	20	74.62	126.85
7	100	30	70	71.07	67.52		70	30	71.07	110.16
8	100	30	70	67.68	64.30		70	30	67.68	104.90
9	100	30	70	64.46	61.24		70	30	64.46	99.91
10	100	40	60	61.39	67.53		60	40	61.39	85.95
		Net present values		772.16	625.78		Net present values		772.16	1304.62

Author's calculation for illustration

The revised NPV for alternative A shows that when assigning a higher weight to the poor rural households, the weighted NPV of the project is lower than the unweighted NPV. In contrast, the weighted NPV for alternative B is higher. If more benefits are likely to accrue to the poor and achieving social equity is a key objective, then alternative B would be preferred.

NPV incorporating distributional weights can be expressed as

$$\text{NPV} = \sum_{j=1}^{m}\left[W_j \sum_{t=1}^{\infty} \frac{b_{t,j} - c_{t,j}}{(1+r)^t} \right]$$

where W_j is the distributional weight for group j; $b_{t,j}$ are the benefits received by group j in period t; $c_{t,j}$ are the costs imposed on group j in period t; m is the number of groups; and r is the social discount rate. In this formula, net benefit of each group is weighted and then aggregated to compute the net social benefit. Such weighted CBA can be applied in assessing policies where distributional issues are of particular concern.

Make a Recommendation

The final step in CBA is to select the best alternative based on the net social benefit. This process needs to consider the counterfactual, that is, the benefit to society in the absence of the project. The recommendation must clearly justify the choice of alternative based on the assumptions made, externalities, risks, uncertainties, and distributional consequences. The choice must also reflect the preferences of all stakeholders, ideally through continuous engagement.

APPLICATIONS OF CBA

CBA, if properly conducted, can be a powerful tool for policy design and decision-making, but its use and effectiveness in project decisions are worth reviewing on an ongoing basis. We find that CBA is in many cases used ex ante to improve decision-making about projects and the choice among alternatives. It is also sometimes used ex post for assessments of the performance of projects.

The World Bank examined how CBA was used in World Bank-supported projects by interviewing 51 project leaders randomly selected from all projects that closed in 2006–2007 and 2008–2009 (IEG 2010b). The interviews revealed that in the experience of more than 80 percent of the project leaders, CBA was usually conducted after the decision had been made to implement a project and that CBA was not the key criterion in deciding whether to fund a project. It was pointed out that the cost-and-benefit data were made available too late to conduct an ex-ante analysis to assist decision-making, which deprived the staff of the needed motivation to conduct a high-quality CBA since the decision of financing the project had already been made.

Annema (2013) reviewed some studies on the use of CBA in megaproject planning. A summary of the findings from the review suggests that political decisions do not necessarily use CBA's main outcomes, such as NPV or benefit-to-cost ratio. One case where CBA was used as a screening device was for developing the Swedish National Transport Investment Plan 2010–2021 (Eliasson and Lundberg 2012). CBA was successfully utilized in investment planning because the Swedish government, recognizing the importance of CBA as a screening tool, conducted CBA at the project preparation stage. The result of the CBA was used in selecting the most viable project from the list of suggestions, leading to an increase in combined benefits. Politicians' initial selection had estimated benefits of 50 billion Swedish kronor. The project selected under CBA criterion has project benefits amounting to 72 billion Swedish kronor.

In some cases, CBA is conducted after the implementation of the project to assess performance. PROGRESA, a large-scale initiative to alleviate poverty in Mexico, was implemented in 1997. The first CBA of PROGRESA was conducted by Coady (2000). All evaluations before that were IEs on outcomes such as reduction in poverty levels or increase in school enrollment.

Coady (2000) mostly conducted cost-effectiveness analysis because of the difficulty in attaching a monetary value to the unpriced benefits thought to arise from educational investments. In 2011, Barham (2011) performed a CBA on the health component of PROGRESA and reported its success based on a benefit-cost ratio, which ranged from 1.3 to 3.6. This benefit-cost ratio was still an underestimation of the total health benefits of PROGRESA, as the benefit estimation only used infant deaths averted.

CBA has also been applied to evaluate institutional policies, such as the implementation of ADB's safeguards policy pertaining to inclusive and environmentally sustainable economic growth. The evaluation estimated the benefits and costs of implementing environmental and involuntary resettlement safeguards on a road rehabilitation project in Sri Lanka (IED 2016). The CBA generated counterfactual scenarios with and without the safeguards. While data were limited, the CBA still generated valuable results. Two out of the three road segments for which the CBA was conducted had positive NPV under the "with safeguards" scenario, and all road segments had negative NPV under the "without safeguards" scenario. The process of conducting this CBA was equally valuable, highlighting methodological challenges (e.g., with respect to assigning values) and data gaps in conducting CBA of safeguards policies.

A CBA of a proposed policy to introduce cleaner fuel in Mexico highlights the application of CBA to environmental protection policies. In Mexico, a country with a severe air pollution problem, a CBA was conducted prior to designing a policy on fuel quality improvement, which allowed policy-makers to incorporate the CBA results into their plan (see Box 3.1). Linked to SDG 11 on sustainable cities and communities and to SDG 13 on climate action, this example, reported in Rojas-Bacho et al. (2013), reiterates the importance of the timing of analysis, as CBA can clearly bring in more benefits if done prior to the beginning of the decision-making process.

Box 3.1 CBA of Fuel Quality Improvement in Mexico (Rojas-Bacho et al. 2013)
Despite the Mexican government's efforts to reduce air pollution by shutting down factories and refineries in Mexico City, air pollution continued to cause approximately seven thousand deaths per year. To reduce the emissions from private vehicles and trucks, which were the main source of pollutants, the Ministry of Environment and Natural Resources (SEMARNAT) revised the fuel quality standard in 2006, aiming to reduce sulfur levels in gasoline and diesel.

Prior to making the policy decision, SEMARNAT and PEMEX (a state-owned fuel producer) were required to conduct a CBA of low-sulfur fuel production. This followed the 2001 government mandate that all federally funded investment projects must carry out a CBA. In 2002, SEMARNAT formed a working group with PEMEX and other public and private institutions, including the National Institute of Ecology (INE), to design the policy. The first proposal, put together in 2003, aimed for sulfur levels in all fuels to be reduced by 2008. This required upgrading refinery infrastructure, which was initially estimated to cost about US$2 billion, a number later revised to US$2.7 billion (Image 3.1).

Image 3.1   Pollution in Mexico City. (Credit: Usfirstgov/CC BY)

INE estimated the benefits and costs over the period 2005–2030. However, it was the first time where a major federal investment project used CBA and included environmental externalities. INE formed a scientific panel to provide advice on conducting CBA. Another challenge was to find reliable estimates of the effects of poor air quality on mortality. As a previous study on Mexico City that was used to compute these estimates only captured short-term effects, INE used the estimates from cohort studies in the United States, which measured long-term effects. The final challenge was related to the monetization of health benefits. There was only one prior study on Mexico on the WTP for mortality risks. It used two methods—hedonic wages, which compared average wages in dangerous jobs with safer jobs to place value on safety, and contingent valuation, which asked how much individuals were willing to pay for a reduction in child mortality risk. INE decided to use values from a meta-analysis of studies in the United States and adjust for Mexican incomes using an estimate of WTP for health with rising income.

The final CBA was submitted to the Ministry of Finance in 2006, and the Congress approved the project. As the supply of ultra-low-sulfur diesel completely depended on imports, compliance proved costly. Domestic technical and engineering plans made no progress either. Consequently, the plan for ultra-low-sulfur diesel had to be delayed by some four years. A revised proposal was made to increase the project's budget to US$5.9 billion, which called for another CBA. Although some benefits were lost due to the delay in compliance and increased costs, the revised CBA showed positive net benefits. After the revised funding was approved by the Ministry of Finance, cleaner fuels became available nationwide by 2011.

The Mexican government realized the need for improving technical capacity and developing a set of guidelines for CBA, including guidelines on discount rates and monetary valuation of intangible benefits.

Another example of CBA applied to environmental policy pertains to water recycling projects in India. The CBA dealt with the externalities generated by small-scale water projects (Labhasetwar 2013; NEERI 2007). It required identification and valuation of unintended consequences such as water pollution and health outcomes. The findings revealed that CBA can make greater contributions if it is done thoroughly, considering all potential externalities, as was done in this study (see Box 3.2).

Box 3.2 CBA of Water Projects in India (Labhasetwar 2013)
Water scarcity in India is serious and worsening. Freshwater availability in India is only 1851 cubic meters per capita per year, compared to an average of 9974 cubic meters per capita per year in the United States (FAO 2012). Surface water availability continues to fall, and the per capita yearly surface water availability in 2050 is projected to be nearly half the amount in 1991 (Kumar et al. 2005). Adding to water scarcity, water quality deterioration imposes costs on human health, the environment, and agricultural production.

India has invested a large amount of money in water-resource projects, whose costs and benefits are long term and go beyond the project site, producing significant uncertainty. CBA can assist in answering questions on the types of water projects to be approved based on calculation of costs and benefits including externalities. Conducting CBA of small-scale water projects involves gathering lot of project-specific scientific, engineering, and economic details. However, with growing knowledge of the technique, there is also wider use of CBA in evaluating small-scale water projects.

In Madhya Pradesh, the infrastructure for proper wastewater disposal was inadequate. A third of rural households and a quarter of urban households had no wastewater drainage system, causing groundwater contamination and making drinking and irrigation water unsafe. There was therefore a need for a cost-efficient gray water (defined as wastewater from showers and basins but not from kitchen or toilet waste) recycling system. A CBA of such a project was conducted by several organizations including the National Environmental Engineering Research Institute (Godfrey et al. 2009) (Image 3.2).

Image 3.2 Gray water reuse facility at secondary school. (Credit: The Sustainable Sanitation Alliance Secretariat/CC BY)

The CBA (see Godfrey et al. 2009) estimated investment costs for gray water treatment and reuse system and for land, civil works and facilities, and piping works as well as any negative externality from gray water reuse. Financial costs (resulting from financing the investment) and operating and maintenance costs of a gray water recycling system were considered on the cost side.

Estimation of benefits from the project included avoided expenses or savings from improved access to water infrastructure as well as health benefits in terms of avoided health expenditures. Benefits also covered environmental impacts such as avoided overexploitation of groundwater, reduced water pollution, and positive spillover on agriculture.

The results (Table 3.5 below) showed that the estimated benefits of a gray water reuse system are far higher than the costs. Based on this finding, the government of Madhya Pradesh allocated funds for the construction of 412 gray water reuse systems in April 2006.

Table 3.5 Summary of annualized cost and benefits (in Indian Rupees)

S. no.	Parameter	Annualized benefit	Annualized cost
1	Capital cost of gray water reuse system		INR 6036 (interest rate @12% per annum)
2	Operation and maintenance cost		INR 5725
3	Availability of gray water	INR 30,000	
4	Avoidance of water infrastructure	INR 50,000	
5	Environmental benefits	INR 44,000	
6	Health benefits	INR 793,380	
7	External cost		Negligible

Source: Godfrey et al. (2009)

Conclusion

CBA is a valuable tool in the evaluator's kit for quantifying the likely net gains from investing in projects and taking certain policy directions. It provides an intuitive and empirical framework to think about both the cost side and the benefit side of interventions and compare the two to make judgments about net benefits. Where well applied, CBA has helped make crucial policy decisions. Importantly, it can be a valuable tool in assessing individual or collective efforts in furthering the SDGs.

The limits to the use of CBA have to do with its scope, which often is within the confines of individual projects and not encompassing broader policy or strategy framework adequately. The availability of data to compute monetary values of social outcomes has also been a constraint. There are several questions on valuation, for example, of items not traded in the market or issues of present versus future value. As with IEs, technical, organizational, and political challenges can also impede CBA from being conducted in a way that is timely and useful.

Increasingly, attention has shifted toward measuring the causal impact of policies, which puts the spotlight on the use of IE. IE and CBA are complementary tools of policy decision-making. While CBA supports initial decisions on investment, it also emphasizes the cost-side assessment, which is often lacking in IE. Without information on costs, there is no means to determine whether an investment is worthwhile.

CBA also has the flexibility to be applied to complex sustainable development issues. For instance, long-term cost and benefit estimation can take into account the environmental impact of a project in the future. Distributional impacts can also be evaluated by assigning appropriate distributional weights, thus giving explicit consideration to issues of equity.

Like IEs using panel data or follow-up of beneficiaries for long periods, CBA can also be conducted more dynamically. As the estimation of costs and benefits involves uncertainty, it is important to monitor the process and review the CBA. Costs and benefits can then be recalculated based on a midterm evaluation.

There is a need for better data collection and improvement in methods so that CBA can be increasingly applied to issues of sustainable development. There is great potential in complementing CBA with IE and OBE to produce high-quality and useful evaluations.

Bibliography

Annema, Jan Anne. 2013. "The Use of CBA in Decision-Making on Mega-Projects: Empirical Evidence." In *International Handbook on Mega-Projects*, edited by Hugo Priemus and Bert van Wee, 291–312. Cheltenham, UK: Edward Elgar.

Barham, Tania. 2011. "A Healthier Start: The Effect of Conditional Cash Transfers on Neonatal and Infant Mortality in Rural Mexico." *Journal of Development Economics* 94 (1):74–85.

Boardman, Anthony E., David H. Greenberg, Aidan R. Vining, and David L. Weimer. 2006. *Cost-Benefit Analysis: Concepts and Practice, Vol. 3*. Upper Saddle River, NJ: Prentice Hall.

Brent, Robert J. 2007. *Applied Cost-Benefit Analysis.* Cheltenham, UK: Edward Elgar.

Chen, Yvonne, Namrata Chindarkar, and Yun Xiao. 2018. Increasing Child Immunization through Uninterrupted Power. *Lee Kuan Yew School of Public Policy.* mimeo.

Coady, David. 2000. The Application of Social Cost-Benefit Analysis to the Evaluation of PROGRESA. Washington, DC: International Food Policy Research Institute.

Eliasson, Jonas, and Mattias Lundberg. 2012. "Do Cost-Benefit Analyses Influence Transport Investment Decisions? Experiences from the Swedish Transport Investment Plan 2010–21." *Transport reviews* 32 (1):29–48.

FAO (Food and Agriculture Organization of the United Nations). 2012. Water Resources: Total Renewable Per Capita (Actual) (m3/inhag/yr), Aquastat Database. Rome: FAO.

Farrow, Scott, and Richard O. Zerbe. 2013. *Principles and Standards for Benefit-Cost Analysis.* Cheltenham, UK: Edward Elgar.

Godfrey, Sam, Pawan Labhasetwar, and Satish Wate. 2009. "Greywater Reuse in Residential Schools in Madhya Pradesh, India: A Case Study of Cost-Benefit Analysis." *Resources, Conservation and Recycling* 53 (5):287–293.

Harberger, Arnold C. 1971. "Three Basic Postulates for Applied Welfare Economics: An Interpretive Essay." *Journal of Economic Literature* 9 (3):785–97.

Harberger, Arnold C., and Glenn Jenkins. 2002. *Cost-Benefit Analysis.* Cheltenham, UK: Edward Elgar.

IED (Independent Evaluation Department). 2016. Real-Time Evaluation of ADB's Safeguard Implementation Experience Based on Selected Case Studies. Manila: ADB.

IEG (Independent Evaluation Group). 2008. The Welfare Impact of Rural Electrification: A Reassessment of the Costs and Benefits. Washington, DC: World Bank.

IEG (Independent Evaluation Group). 2010a. Assessing the Impact of IFC's China Utility-Base Energy Efficiency Finance Program. Washington, DC: World Bank.

IEG (Independent Evaluation Group). 2010b. Cost-Benefit Analysis in World Bank Projects. Washington, DC: World Bank.

Jenkins, Glenn P., and Arnold C. Harberger. 1998. *Cost-Benefit Analysis of Investment Decisions: Manual.* Cambridge, MA: Harvard Institute for International Development.

Jeuland, M. A. , V. Bhojvaid, A. Kar, J. J. Lewis, O. Patange, S. K. Pattanayak, N. Ramanathan, I. H. Rehman, J. S. Tan Soo, and V. Ramanathan. 2015. "Preferences for Improved Cook Stoves: Evidence from Rural Villages in North India." *Energy Economics* 52:287–298.

Kind, Jarl, W. J. Wouter Botzen, and Jeroen C. J. H. Aerts. 2017. "Accounting for Risk Aversion, Income Distribution and Social Welfare in Cost-Benefit Analysis for Flood Risk Management." *Wiley Interdisciplinary Reviews: Climate Change* 8 (2):e446.

Kumar, Rakesh, R.D. Singh, and K.D. Sharma. 2005. "Water Resources of India." *Current science* 89 (5):794–811.

Labhasetwar, Pawan. 2013. "Cost-Benefit Analysis of Water Projects in India." In *The Globalization of Cost-Benefit Analysis in Environmental Policy*, edited by Michael A. Livermore and Richard L. Revesz. New York: Oxford University Press.

Layard, P. Richard G. 1994. *Cost-Benefit Analysis*. Cambridge: Cambridge University Press.

Mahasuweerachai, Phumsith, and Siwarut Pangjai. 2018. "Does Piped Water Improve Happiness? A Case from Asian Rural Communities." *Journal of Happiness Studies* 19 (5):1329–1346.

Mulder, Peter, and Jonas Tembe. 2008. "Rural Electrification in an Imperfect World: A Case Study from Mozambique." *Energy Policy* 36 (8):2785–2794.

National Environmental Engineering Research Institute (NEERI). 2007. Greywater Reuse in Schools – A Guidance Manual. Nagpur: NEERI.

Nordhaus, William D. 2007. "A Review of the Stern Review on the Economics of Climate Change." *Journal of Economic Literature* 45 (3):686–702.

Pearce, David, Giles Atkinson, and Susana Mourato. 2006. *Cost-Benefit Analysis and the Environment: Recent Developments*. Paris: OECD.

Petri, Peter, and Vinod Thomas. 2013. Development Imperatives for the Asian Century. *ADB Economics Working Paper Series no. 360*. Manila: ADB.

Ray, Anandarup. 1984. *Cost-Benefit Analysis: Issues and Methodologies*. Baltimore: Johns Hopkins University Press.

Rojas-Bacho, Leonora, Veronica Garibay-Bravo, Gretchen A. Stevens, and Georgina Echaniz-Pellicer. 2013. "Environmental Fuel Quality Improvements in Mexico: A Case Study of the Role of Cost-Benefit Analysis in the Decision-Making Process." In *The Globalization of Cost-Benefit Analysis in Environmental Policy*, edited by Michael A. Livermore and Richard L. Revesz. New York: Oxford University Press.

Sinden, John Alfred, and Dodo J. Thampapillai. 1995. *Introduction to Benefit-Cost Analysis*. Melbourne: Longman

Squire, Lyn, and Herman G. Van der Tak. 1975. *Economic Analysis of Projects*. Washington, DC: World Bank.

Stern, Nicholas Herbert. 2006. *The Economics of Climate Change: The Stern Review* Cambridge: Cambridge University Press.

Tan Soo, Jie-Sheng. 2017. "Valuing Air Quality in Indonesia Using Households' Locational Choices." *Environmental and Resource Economics* 1–22.

Tietenberg, Thomas H., and Lynne Lewis. 2016. *Environmental and Natural Resource Economics*. Oxton and New York: Routledge.

Tolley, George S., and Robert G. Fabian. 1998. "Issues in Improvement of the Valuation of Non-Market Goods." *Resource and Energy Economics* 20 (2):75–83.

Whittington, Dale, and Duncan Macrae. 1986. "The Issue of Standing in Cost-Benefit Analysis." *Journal of Policy Analysis and Management* 5 (4):665–682.

World Bank. 2000. Energy Services for the World's Poor. Washington, DC: World Bank.

World Bank. 2018. Tracking SDG7: The Energy Progress Report 2018. Washington, DC: International Bank for Reconstruction and Development, World Bank.

Zerbe, Richard O., and Allen S. Bellas. 2006. *A Primer for Benefit-Cost Analysis.* Cheltenham, UK: Edward Elgar Publishing.

Zhuang, Juzhong, Zhihong Liang, Tun Lin, and Franklin De Guzman. 2007. Theory and Practice in the Choice of Social Discount Rate for Cost-Benefit Analysis: A Survey. *ERD Working Paper Series no. 94.* Manila: ADB.

Open Access This chapter is licensed under the terms of the Creative Commons Attribution 4.0 International License (http://creativecommons.org/licenses/by/4.0/), which permits use, sharing, adaptation, distribution and reproduction in any medium or format, as long as you give appropriate credit to the original author(s) and the source, provide a link to the Creative Commons licence and indicate if changes were made.

The images or other third party material in this chapter are included in the chapter's Creative Commons licence, unless indicated otherwise in a credit line to the material. If material is not included in the chapter's Creative Commons licence and your intended use is not permitted by statutory regulation or exceeds the permitted use, you will need to obtain permission directly from the copyright holder.

CHAPTER 4

Objectives-Based Evaluation for Accountability and Learning

Abstract This chapter features objectives-based evaluation (OBE) as a means to assessing the performance of international development finance, which remains key to achieving sustainable development priorities in many parts of the world. It lays out key evaluation criteria included in OBE and illustrates its multi-level approach that not only focuses on project-level outcomes but also on aggregate-level outcomes such as sectoral, country, or thematic level. It emphasizes the critical role of OBE in providing a big picture of how well development resources are being utilized. Again, it pays special attention to the use of the method with respect to issues of sustainable development.

Keywords Development finance • Accountability • Learning • Effectiveness • Efficiency • Sustainability

Commitments of development finance naturally bring an emphasis on the delivery of agreed objectives. Where there is a stress on accountability in delivering on agreed goals, objectives-based evaluation (OBE) assumes significance. Equally, the lessons from such assessments inform what works and what does not.

Complementing inputs from impact evaluation (IE) and cost-benefit analysis (CBA), the OBE approach is applied to conduct individual project or program assessments and also aggregate sector- or country-level assessments. The assessment of the Sustainable Development Goals (SDGs) can occur at the micro project levels and also at the more macro and aggregative levels, as OBEs can ask about the extent to which the different interventions taken together address overall directions, for example, the pursuit of the SDGs.

Objectives-Based Evaluation

Overseas development assistance (ODA) is one of the core resources developing countries can use to work toward achieving the SDGs. ODA recipients are low- to middle-income countries who use the funds to invest in a wide range of sectors. The amount of ODA has increased over the past few years, exceeding US$150 billion in 2015 (at 2015 constant prices, according to OECD.Stat). At the same time, the size of ODA has been eclipsed by the size of the development financing labeled as coming from the private sector.

As development finance is used in more diversified ways, there is increasing recognition that it should be provided based on harmonized objectives and accountability. Harmonizing objectives can help improve efficiency in the use of resources, while accountability of both donors and recipients sheds light on how results are being achieved (or not) by development finance. OBE can facilitate both harmonization and accountability by establishing a set of objectives and examining achievements against these stated objectives. Assessing the effectiveness of development finance generates information for accountability and learning concerning the outcomes.

OBE is a class of evaluation methods that considers the extent to which objectives are achieved. Its strong focus on measuring outcomes against initially stated objectives encourages the formulation of clear objectives from the outset. It is guided by a theory of change that connects the inputs and activities to the outcomes. As opposed to output targets, which are set in terms of the tangible goods and services that project activities produce, outcomes measure the results likely to be achieved once the beneficiary population starts using the project outputs. Measures of outcomes look at the final results achieved, which indicate whether project goals are met (Gertler et al. 2016).

Output is something that can be controlled by a project-implementing agency. In a water project, for example, output can be measured by the number of pipes constructed. In contrast, outcomes can take longer to manifest and be observed long after the project has been completed. For instance, improved child health, measured using height-for-age (stunting) z-scores, is a potential outcome of piped water connections.

Different outcomes may be realized on different time horizons: the construction of pipes may save time and make collecting water more convenient for beneficiary households in the short-to-medium time frame, while in the longer term it may lead to improved health.

OBE Criteria

The standard criteria used in OBE are relevance, effectiveness, efficiency, sustainability, and considerations of development impact. While specified separately, these criteria also have a great deal of connectivity among them. More broadly, assessments of individual projects or aggregate programs are related to considerations of social inclusion, environmental protection, and governance, but at present these development goals are not directly used as evaluative lenses. Ratings for the standard criteria commonly use a four- or six-point scale, with categories that differ by rating institution. As an example, ratings used by the World Bank and Asian Development Bank (ADB) are summarized in Table 4.1 (IED 2016; IEG 2017a).

Table 4.1 Criteria and ratings of OBE: example of the World Bank and ADB

	World Bank	ADB
Overall criteria	Highly satisfactory; satisfactory; moderately satisfactory; moderately unsatisfactory; unsatisfactory; highly unsatisfactory[a]	Highly successful; successful; less than successful; unsuccessful[b]
Relevance	High; substantial; modest; negligible	Highly relevant; relevant; less than relevant; irrelevant
Efficacy/ effectiveness	High; substantial; modest; negligible	Highly effective; effective; less than effective; ineffective
Efficiency	High; substantial; modest; negligible	Highly efficient; efficient; less than efficient; inefficient
Sustainability	Most likely; likely; less than likely; unlikely	Most likely sustainable; likely sustainable; less than likely sustainable; unlikely sustainable
Development impact	Highly satisfactory; satisfactory; less than satisfactory; unsatisfactory	Highly satisfactory; satisfactory; less than satisfactory; unsatisfactory

[a]Composite rating of core criteria—relevance, efficacy, and efficiency
[b]Weighted average of core criteria—relevance, effectiveness, efficiency, and sustainability

Relevance assesses the extent to which the financed activity is suited to a country's development priorities. It asks whether the objectives of the project are valid and whether the activities, outputs, and project design are consistent with the objectives. *Relevance* considers analyses and lessons on project design, including the financing instrument and modality, as they relate to the objectives. It is important to consider both what is included in the project and what ought to be included. This criterion also looks into how the project takes into account the work of development partners and other organizations. All this means that a project, by the fact that it is approved, does not automatically qualify as being relevant; the evaluator must assess if it is.

Effectiveness measures the extent to which the project's intended outcomes are achieved, based on a baseline and a target. For a project to be effective, outcomes should have been achieved or be likely to be achieved, and output targets should also have been substantially achieved. It also looks at factors influencing the achievement of the objectives. This is important for assessing the extent to which the outcomes achieved were a contribution of the project's interventions. Achieved outcomes need to be plausibly attributable to the project or intervention as established through either IE or a theory-based approach and consideration of the counterfactual. If outcome targets were achieved or are likely to be achieved but output targets were not substantially met, the project will not be considered effective. This is because some outcomes at the national level, such as a higher literacy rate, may be achieved by other interventions than the one being evaluated.

Where projects have multiple objectives and outcomes, the evaluation can assign relative weights to the effectiveness of various objectives and outcomes. For example, a project may be implemented in two distinct geographical areas with one having a higher incidence of poverty. If intervention in the poorer area is seen as a higher priority, the poverty outcome in that area might be given more weight.

Efficiency measures how well resources are used to achieve outcomes. It assesses the project's economic benefits against economic costs using CBA. In assessing efficiency, CBA, in principle, looks at costs and benefits, including unintended and indirect ones, from a societal viewpoint. Project economic-performance indicators—the economic internal rate of return, net present value, and the cost-benefit ratio—are used to determine whether net gains from investing in a project will be enjoyed by the society following project completion.

In many instances, it is instructive to compare alternative approaches to achieving the same results using a least-cost analysis. The cost estimates should be based on the economic costs incurred to implement the project, as well as provisions for the operation and maintenance of the assets over the expected economic life of the project. Project externalities—which are spillover effects and often unintended consequences such as environmental impacts—should be quantified and valued to the extent possible and incorporated into the calculations.

Sustainability here focuses on the more limited aspect (compared to the broad goal of sustainable development) of the likelihood that an activity will be maintained after donor funding has been withdrawn. It considers the major factors influencing the achievement or non-achievement of sustainability of the project's results. Since an evaluation is typically carried out during the first few years of a project's operational life, evaluators must make assumptions about the likely sustainability of operational arrangements, many of which are new, and about probable future operations and maintenance (O&M) arrangements. Even within this narrower confine of just one of the evaluation criteria, however, the environmental effects of a project must also be considered such as the effects on natural resource management, pollution, biodiversity, and greenhouse gas emissions.

This emphasis on environmental protection and climate change mitigation or adaptation is a relatively new requirement in evaluating projects. Climate mitigation is crucial, as the cost of adaptation is high. The less done to mitigate climate change, the more severe and expensive are the consequences. Assessments of sustainability should consider political, economic, institutional, technical, social, environmental, and financial risks. The assessment should also consider the adequacy of risk-mitigation measures. Although environmental- and social-risk avoidance may be part of the effectiveness assessment at some institutions, grave social or environmental consequences of failed mitigation may also affect the sustainability assessment.

Development impact is the positive and negative change generated by a development intervention, directly or indirectly, intended or unintended, and it mirrors the degree to which sustainable development is being achieved. It assesses the overall changes attributable to the intervention, the difference made to the beneficiaries, and the number of people affected. Development impacts to which the project contributes tend to be outside the project's direct control, and their achievement is often not solely attributable to the project outcomes. Typically, they are dependent on other development efforts. The focus of analysis should be on the contribution of project outcomes to the achievement of the project impacts. Moreover, development impacts can also be due to unforeseen events and positive developments in areas that are outside the project scope. Such impacts should not be attributed to the project.

Multilateral development agencies have been using these criteria since the 1990s. In 1991, the Organisation for Economic Co-operation and Development-Development Assistance Committee (OECD-DAC) issued principles for evaluation of development assistance provided to public sector projects. The principles provided general guidance on evaluating projects financed using development assistance, stating that the aim of evaluation was "to determine the relevance and fulfilment of objectives, developmental efficiency, effectiveness, impact and sustainability" (OECD-DAC 1991). Also emphasized was the importance of formulating the objectives to be achieved at the baseline for high-quality OBEs.

Aggregative evaluations also incorporate these criteria. In response to the growing importance of harmonized evaluations, a workshop was held by the Working Group on Aid of the OECD-DAC on country-program evaluation methodology in 1999, and it was agreed that the criteria for evaluating development finance at the country level could draw on these criteria. Since the 1999 workshop, a common core of good practices for country evaluations has evolved, and gradually these common criteria have been adopted and endorsed by the independent evaluation offices of various Multilateral Development Banks (MDBs).

Project Evaluations

OBE assesses the outcome of development finance not only at the individual project level but also at aggregate levels such as county, sector, or theme. Lessons across projects from such aggregative assessments can bring good value.

That said, project evaluation is not only an in-depth assessment of what occurred in the investment but also a vital basis for what then can be surmised at the thematic or country levels. This means there is more to be learned from clusters of projects. Project-level evaluations are required as

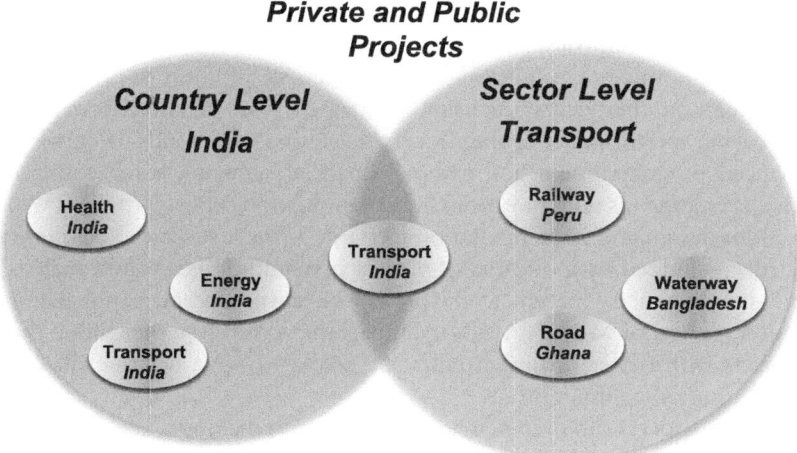

Fig. 4.1 OBE of development finance. (Source: Authors' illustration)

important building blocks to higher-level evaluations. Individual projects, for example, in transport or health, are a primary focus of OBE providing evidence on how development financing channeled through them are achieving the envisaged objectives.

As illustrated in Fig. 4.1, these project results are also building blocks for aggregative assessments spanning sectors like education or energy; themes like urban development or the environment; or nations, as in country assessments for China or Sri Lanka. They can also be inputs for corporate studies of how an organization is doing in different aspects of its development work.

Public Sector Projects

Much of the OBE work is concerned with the performance of individual projects designed and implemented by governments. Applying the criteria discussed earlier, assessments arrive at conclusions on project success or failure in varying degrees. These success rates also vary a great deal across sectors, across countries, and over time.

The main value of delineating success rates (in addition to providing a judgment on the success in using financing) is in deriving lessons for improving performance going forward. Across thousands of projects financed by government and external financial agencies, some common lessons emerge (as seen in various annual evaluation reports from MDBs).

Both design and implementation feed into project success. The *quality at entry*—judged by indicators of project preparation, economic analysis, and due diligence—matters to the project's eventual success. Annual evaluation reports of MDBs provide examples that taken together signal that projects with better quality at entry are seen to have a better chance of succeeding on completion.

Several aspects of implementation stand out in making a key difference to project success. Adequate allocation of budget for implementation and allowing for adequate spending on operations and maintenance of infrastructure investments are obvious considerations. The importance of pricing and regulatory policy frameworks is clear in the examples of water pricing or energy regulation. Institutional capacities of implementing countries are also important for project success.

Historically, project success in low-income countries has been seen to be lower than in middle-income countries, but this pattern is by no means universal. Furthermore, this gap has narrowed over time, as seen in the annual evaluation reports of Asian Development Bank and the World Bank. East Asia has been a perennial front-runner when it comes to project outcomes, a result especially driven by China's strong project performance. Historically, infrastructure (energy in particular) has done well, though social sectors, environment, and multi-sectoral investments have strengthened in many settings over time.

The case of a water supply and sanitation project in Sri Lanka illustrates how these criteria are applied (IEG 2017b). There was uneven access to water and sanitation between urban and rural areas. Therefore, a water supply and sanitation program was implemented in 2004 with the objectives of increasing coverage and achieving effective and sustained use of water and sanitation services in rural communities. The project's aim would correspond directly with what SDG 6 set out in 2015 for clean water and sanitation. The project was approved in 2003 and closed in 2010. The total costs were estimated to be US$69.1 million.

To prepare the project performance assessment report (PPAR), seven visits to project sites were made. In-depth discussions among focus groups were conducted to gather information to assess the project outcomes. The development outcome of the project was rated *moderately satisfactory* based on an assessment of *relevance, efficacy,* and *efficiency* (Table 4.2).

Relevance of the development objective was rated *substantial* as it was aligned with both the government's priorities and with the World Bank's country assistance strategies with respect to expanding water and sanitation services to the rural population. Project design relevance was also

Table 4.2 Assessment of community water supply and sanitation project in Sri Lanka

Criteria	Ratings	Assessment
Overall development outcome	Moderately satisfactory	Based on the individual ratings below
Relevance of development objective	Substantial	Project objectives were relevant to both the government's priority and the World Bank's country strategy
Relevance of project design	Substantial	The main components and results framework were generally clear and logically linked to the project's objectives
Efficiency	Substantial	ERR and NPV were estimated for the two most popular piped water supply technologies (piped gravity and pumping schemes): the resulting ERR was 30% for gravity and 18% for pumping schemes at completion. NPV per household was SL Rs 11,000 for gravity and SL Rs 2000 for pumping schemes at completion
Outcome of objective 1: to increase service coverage	Modest	Modest achievement of key indicators, such as the number of people provided with access to improved water sources and new piped household water connections established
Outcome of objective 2: to achieve effective and sustained use[a]	Substantial	Some challenges in ensuring reliability and water quality (such as lack of 24-hour supply and water contamination). Overall, the project provided adequate, affordable, and relatively sustainable water services, ensuring convenience and time saving for the beneficiaries

Source: IEG (2017b)
[a]Sustainability is assessed within objective 2

rated *substantial*. The main components of the project design and its results framework were clear and logically linked to the project's objectives of increasing service coverage and achieving effective and sustained use of water and sanitation services in rural communities.

The evaluation assessed *efficacy*, or the achievement of project objectives of service delivery and management, as well as demand and sustainable water use. Specifically it considered (1) increasing service coverage and (2) achieving effective and sustained use of water and sanitation services in rural communities in Sri Lanka.

The outcome of the first objective under *efficacy* was rated *modest*. The original target was not achieved. The number of people provided with access to improved water sources under the project increased by 384,100, but that was only 31 percent of the original target and 48.4 percent of the revised target at completion. The report points to two major reasons: decreased funds owing to the tsunami in 2004 and inflation. The target was not revised sufficiently despite the reduction in funds.

The second *efficacy* objective, which aimed to achieve effective and sustained water use, was assessed using demand-side outcomes: satisfaction, adequacy, reliability, convenience, and time saving, water quality, affordability, and sustainability of services. The achievement of this second objective was rated *substantial*. Satisfaction was generally high, with 88 percent of the beneficiaries of the completed water subprojects indicating they were satisfied with their access and that water adequacy had improved. A survey indicated that household connections led to sizable time savings.

The household survey at completion indicated that about 46 percent of the subprojects provided continuous water supply and 78 percent of households received piped water every day. The Ministry of Health's sampling tests showed that only 44 percent of the subprojects provided water of satisfactory quality. The project achieved relatively high sustainability, but there were cases of water resource depletion and issues of frequent repair needs due to poor technical design in the initial phase.

Efficiency was rated *substantial*. The project provided access to improved water sources to fewer people compared to the original target, with slightly higher costs than estimated at appraisal. But the economic rate of return computed was still relatively high.

Community-based organizations were the main providers of water supply schemes in rural areas and responsible for operation and maintenance of water supply facilities in villages. Therefore, beyond the core criteria, *risk to development outcome* was rated *significant* because many community-based organizations faced technical, financial, and organizational challenges in sustaining the project results. Technical challenges included repair of pumps and water contamination. Furthermore, only a few community-based organizations were seen to be financially sustainable. Finally, some of them suffered from a shortage of volunteers, and the institutional arrangement for supporting them was unclear.

Private Sector Projects

Analogous to public sector projects, private sector projects can be evaluated using OBE. Though some OBE criteria are used for both public and private sector projects, there are important differences in their evaluation. Evaluations of private sector projects focus on business performance, which assesses the effect of the project on its financiers, and economic contribution, which assesses its economic effects more broadly.

Private sector project evaluation, while paralleling that of public sector projects, also has distinct features to account for the public support through MDBs to private businesses (see EBRD 2013; MDBs 2018). Since support to private sector projects is part of the broader activities of MDBs, the project assessment works within the framework provided by the MDBs' overall mission to promote sustainable private sector investment in developing countries while also achieving social outcomes such as inclusive growth and poverty reduction.

Some MDBs have shifted away from methodologies largely driven by business performance. New approaches are based on market benchmarks and apply OBE criteria such as relevance, effectiveness, efficiency, and sustainability to private sector operations. *Relevance* and *effectiveness* in this context focus on both development and business outcomes. *Efficiency* looks at the financial performance (including achievement of business objectives) and economic performance.

An OBE considers a private sector project against its development outcomes, including business performance. The development impact rating is a synthesis of the impact of the project on the country's economic and social development. Development impacts are evaluated using a with-versus-without-project comparison, that is, considering (1) what happened with the project and (2), counterfactually, what would have happened without it. When a with-versus-without assessment cannot be done, the costs and benefits to the country can be done on a before-versus-after basis.

At the World Bank Group, a private project's development outcome is measured across four indicators: project business performance; economic sustainability; environmental, social, health, and safety effects; and contribution to private sector development. Each of these measures a distinct aspect of the project's performance. The development outcome rating is a bottom-line assessment of the project's results on the ground, and not an "average" of these four indicators. Each of the categories can be rated on a four- or six-point scale indicating the degree of success.

A project's *business performance* measures the project's impact on profitability and viability, the project's contribution to other business goals, and the project company's prospects for sustainability and growth. Sufficient financial returns are necessary to attract and reward private investment, but the assessment should also take into consideration the sustainability of the results. Projects can be structured as either loan or equity and can include institution-building components.

Interestingly, previous empirical analysis has found a strong connection between project development outcomes and an organization's financial profitability (IEG 2009). For example, in a cohort of projects financed by the International Finance Corporation (IFC), high/high outcomes (high development outcome and high IFC investment return) were achieved in 66 percent of projects (by the number of projects, not volumes), and another 17 percent had low/low outcomes. This is suggestive of the possibility that focusing on development outcomes is also good business (IEG 2009).

Economic sustainability reflects the project's contribution to economic growth. Projects with high economic returns clearly contribute to a country's economic growth, and this contribution is quantifiable to some extent. Harder to achieve, but vital, is the effort to reduce poverty and improve people's lives. It is important to address to what extent, as a result of a project, resources are being allocated more efficiently and the project portfolio is providing a net economic benefit, including the broader attributes of well-being.

It is important also to assess a project's impact on people other than the investors (or adopt a stakeholder perspective) such as client companies and their customers, employees, government, competitors, and local residents. Examples of economic benefits and costs accruing to them are ease of access to markets and services, greater market efficiency, contribution to government revenues, contribution to poverty alleviation, social or gender equality, and employment generation.

Social and environmental protection are important components of the development outcome of private projects. That operations are carried out in an environmentally and socially responsible manner is not only sound business practice, but it is also a necessary condition for sustainable development. IFC's Expanded Project Supervision Report assesses the project's environmental performance in meeting regulatory requirements as well as the project's environmental impacts through its subprojects, including pollution loads; conservation of biodiversity and natural resources; and social, cultural, and community health aspects, as well as labor and working conditions and workers' health and safety.

Compliance with specific environmental requirements should be clearly stated in the OBE of private projects. Environmental requirements help enhance environmental management capacity and produce sound development outcomes and can be considered as proxy for acceptable environmental standards. But the effects on the ground should count most in evaluating development outcomes.

Supporting *private sector development* or encouraging the growth of private enterprises is a principal goal. Projects need to help create conditions conducive to the flow of private capital into productive investment. It is crucial that the benefits of growth of productive private enterprise accrue to the entire society. Projects can contribute to this purpose by contributing to the growth of sustainable and viable institutions, contributing to the development of the markets in which they operate, and by financing sustainable and viable private enterprises in the real sector.

AGGREGATIVE EVALUATIONS

While OBE of individual projects generates information needed to assess the outcomes of each project, aggregation of project findings into an assessment of a country, sector, or theme can be of great value in seeing the combined effects. In moving from individual project assessment to aggregative evaluations, a key aspect is that projects are not usually independent of each other, but they complement, and sometimes substitute, each other. Projects are, or ought to be, connected to each other through a strategic framework that involves considerations of sequencing, prioritizing, and spillover effects.

When aggregative assessment is attempted, the core criteria are sometimes not weighted, and sometimes weighted according to policy priorities. Each rating uses a point scale of four or six categories going from high to low (e.g., gradations going from highly relevant to irrelevant). The core assessments are usually complementary and interrelated, for example, aspects of efficiency, such as good financial management, complementing effectiveness. At the same time, to avoid double counting, the same factor, say the lack of pricing for a service, may not be used the same way as the contributor to rating two separate criteria, efficiency and effectiveness.

Sector Evaluations

When individual project evaluations are added up to the sector level, as in transport or education, or at the thematic level, such as for the environment or urbanization, they bring out the aggregative impact of various projects. These also help uncover the effect of policies and investments that cut across the sector or theme. For example, a new decentralization policy might affect the collective success rates of public sector projects. Aggregative evaluations also often account for the influence of factors outside of the theme and the sector. For instance, an increase in infrastructure investments can affect the collective performance of education or health projects.

Sector evaluation is a useful tool to assess impact that goes beyond individual projects. Various organizations have conducted a range of sector evaluations, for example, in agriculture (IOE 2017; OVE 2015b), finance (OED 2018), and energy (IDEV 2015).

An example is the evaluation of World Bank-funded projects in education (IEG 2006; see also IFC 2014). This evaluation was based on a synthesis of outcomes of projects that focused on education access and learning, complemented by findings on policies and investments across the sector. The evaluation of individual projects conducted over several years was summarized and then used, along with sector-wide findings, to assess impact at the sectoral level.

The sector evaluation was done on the five criteria discussed earlier. The ratings in Table 4.3 comprise three criteria: *development outcome* (which is a composite rating of *relevance, efficiency,* and *effectiveness*), *sustainability,* and *development impact*.

Given the rising need to shift focus away from merely providing access to schools and toward improving the quality of education and accelerating learning, outcomes were evaluated by objectives related to access as well as learning and quality of education. Table 4.4 suggests that support for

Table 4.3 IEG ratings of completed primary-education projects by year of approval

	Fiscal year approved			All primary-education projects	All education projects excluding primary	All World Bank-supported projects
	Before 1990	1990–1994	1995–1999			
Outcome (% of projects rated moderately satisfactory or better)	76	89	85	82	78	72
Sustainability (% of projects rated likely or highly likely)	50	66	76	62	66	50
Institutional development impact (% of projects rated substantial or high)	20	19	38	25	46	36

Source: IEG (2006, p. 20)
Note: Refer to Table 4.1 for the criteria and rating scale

Table 4.4 Outcomes by enrollment objective for complete primary-education projects

		Fulfillment of objective (percent; N = 20)			
Objective	Number covering objective	Fulfilled	Partially fulfilled	Unfulfilled	Undetermined
Increased enrollment	13	69	0	23	8
Improved equity	12	75	25	0	0
Improved access for girls	9	55	22	22	0
Improved internal efficiency	12	25	42	25	8

Source: IEG (2006, p. 24)

Table 4.5 Outcomes by objective for complete primary-education projects

Objective	Number covering objective	Fulfillment of objective (percent; N = 20)			
		Fulfilled	Partially fulfilled	Unfulfilled	Undetermined
Improved learning outcomes	6	67	17	0	17
Improved educational quality	18	39	27	33	0

Source: IEG (2006, p. 32)

access to primary education in some respects did well, with serious gaps in improved access for girls.

Further, Table 4.5 indicates gaps in improving learning outcomes and improved educational quality.

Another example can be drawn from an evaluation of World Bank support for transport from 1995 to 2005 (IEG 2007). During this period, there were 642 projects with transport components, carrying a total financial commitment of US$32 billion. This was the first evaluation of the transport sector operation of the World Bank, and it provided insights that could not be obtained by evaluating individual projects given the significant diversity within the sector.

This sector-level evaluation also adopted the five criteria as project-level evaluation: *outcome*, which is a composite rating of *relevance, effectiveness,* and *efficiency; institutional development impact;* and *sustainability.* Within the evaluation period, performance of the projects improved over time across all indicators (Table 4.6). Ratings were lower when large borrowers were excluded because bigger countries in this sample had better institutional capacities.

To gain further insight, sector evaluations can be disaggregated by region. This helps evaluate which region for a particular sector might be lagging and perhaps require further financing and technical assistance. For example, Fig. 4.2 shows that in the transport sector, South Asia is lagging in terms of outcome, while the Europe and Central Asia region and the Middle East and North Africa region may require support in institutional development.

Table 4.6 IEG ratings of projects by exit year, fiscal 1992–2006 (transport sector projects versus all others)

IEG rating			Fiscal 1992–1994	Fiscal 1995–1997	Fiscal 1998–2000	Fiscal 2001–2003	Fiscal 2004–2006
Outcome: Moderately satisfactory or better (%)	Transport	All projects	69	78	84	86	89
		Excluding large borrowers	71	70	82	74	88
	All other	All projects	64	67	68	72	79
Institutional development: Substantial or better (%)	Transport	All projects	25	37	63	68	57
		Excluding large borrowers	27	33	57	66	50
	All other	All projects	30	32	37	45	57
Sustainability: Likely or better (%)	Transport	All projects	46	55	70	74	78
		Excluding large borrowers	47	43	66	71	71
	All other		44	47	54	64	71

Source: IEG (2007, p. 22)
Note: See Table 4.1 for the criteria and rating scale

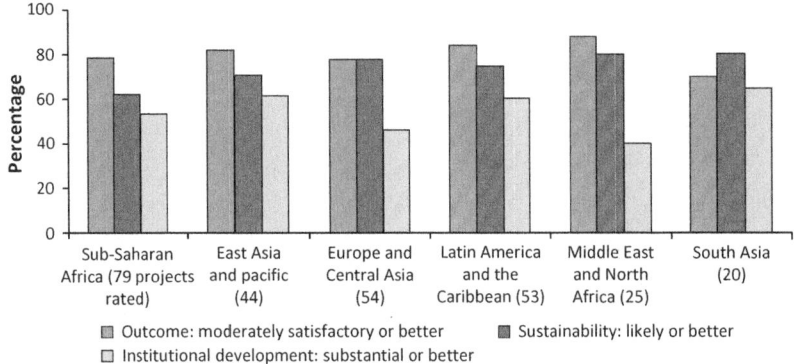

Fig. 4.2 IEG ratings for outcome, sustainability, and institutional development (approval year fiscal 1995–2006). (Source: IEG 2007)

Country Evaluations

Country-level evaluation looks at the nationwide impacts of projects and associated activities of strategy and policy exchanges in a country. Sometimes the success of individual projects does not translate into better country performance. And despite project-level weaknesses, country-level dialogue and impact on sustainable development can be disproportionately high. The role of country expertise, quality of policy discussions, interactions with other partners, and the role played by knowledge are all factors determining country-level results.

For large lenders such as the World Bank and the IMF, the country-level program considerations are especially important, whereas for smaller bilateral financiers, country-level impact may not be that large. A variety of evaluations of country and global macrocosmic work is carried out at the IMF (see, e.g., IEO of the IMF 2016) and the United Nations (for instance, IEO of the UNDP 2018).

A country evaluation of India done in 2017 (IED 2017) assessed the performance of ADB strategy and programs for India from 2007 to 2015. ADB provided US$22.1 billion in development loans to India over these eight years, mostly for transport, energy, finance, and water and other infrastructure and services.

A full-fledged OBE approach for a country evaluation would assess the strategic objectives set out at the beginning of a country strategy and to what extent the objectives were achieved. India's strategic directions present a very broad canvas, and ADB's presence in the financing in terms of size and scope for this large economy is relatively small. Accordingly, the framework for the country evaluation was modest.

The OBE considered the overall performance of development finance to India to be *successful*. It also examined the country's strategic agendas and special priorities, rating them individually. In accordance with India's recent development plan, ADB country strategies for India shifted toward supporting faster, sustainable, and more inclusive growth. It evaluated ADB support by sector as well as support for strategic agendas.

Table 4.7 provides the assessment for the main sectors. The transport program was the largest component in ADB's sovereign portfolio (35 percent) with a value of US$6 billion. Most of the loan went to construction, upgrading, and rehabilitation of rural roads mostly in lagging states in northern and eastern India, which is consistent with the sector strategy to promote connectivity and increase access of rural areas to towns and cities. *Development impacts* of the road program were rated *satisfactory* despite some delays that occurred due to limited capacity of executing agencies and contractors, resulting in *efficiency* being rated *less than efficient*.

In energy, the second-largest component of ADB's sovereign portfolio (29 percent), 33 projects were approved with US$4.9 billion in value. Supporting the government's strategy to develop a strong national grid and expand and optimize transmission and distribution systems, ADB's support focused heavily on electricity transmission and distribution (81

Table 4.7 Sector-wise rating in country evaluation

Sector	Relevance	Effectiveness	Efficiency	Sustainability	Development impacts
Transport	Relevant	Effective	Less than efficient	Likely sustainable	Satisfactory
Energy	Relevant	Effective	Less than efficient	Likely sustainable	Satisfactory
Water and urban infrastructure and service	Relevant	Effective	Less than efficient	Less than likely sustainable	Satisfactory
Finance	Relevant	Effective	Less than efficient	Likely sustainable	Satisfactory
Public sector management	Relevant	Effective	Efficient	Likely sustainable	Satisfactory

Note: Agriculture and natural resource sector and social sector are reviewed without rating percent of the energy sector investment). The energy programs also experienced delays for which *efficiency* was rated *less than efficient*. *Development impacts*, however, were rated *satisfactory* since all completed projects met their objectives of strengthening the transmission system, reducing losses, improving evacuation capacity and system reliability, and increasing access to electricity.

Water and urban infrastructure and services, finance, and public sector management were considered to have achieved *satisfactory development impacts*. Accordingly, the evaluation considered the overall achievement of these sectors to be *successful*.

Table 4.8 shows that the report also assessed ADB's support for three primary strategic agendas: inclusive economic growth, environmentally sustainable growth, and regional cooperation and integration. The report also assessed special priorities: knowledge solutions and capacity development, gender equality, and catalyzing infrastructure investment and public-private partnerships.

Among the three strategic agendas, inclusive growth was most prominent in ADB's lending portfolio to India, with some 90 percent of projects intended to contribute directly or indirectly to this objective. This was accomplished most notably through the financing that was made available to economically lagging states. In addition, ADB also supported inclusive growth through incorporating inclusive elements into project design (for instance, slum improvement in urban projects) and ensuring the

Table 4.8 Assessment of the country's strategic agendas and special priorities

	Relevance	Development impacts
Inclusive economic growth	Relevant	Satisfactory
Environmentally sustainable growth	Relevant	Satisfactory
Regional cooperation and integration	Relevant	Less than satisfactory
Knowledge and capacity development	Less than relevant	Satisfactory
Gender equality	Relevant	Satisfactory
Catalyzing infrastructure investment and public-private partnerships	Relevant	Satisfactory

Source: IED (2017)

implementation of projects in sectors with high inclusion impact, such as education, water supply, and rural electrification. About half of ADB's support program was tagged for environmentally sustainable growth, effected mainly through projects in energy and water and urban infrastructure and services.

A similar country evaluation of development assistance to Brazil was done by the World Bank (IEG 2015; see also OVE 2015a). The support aimed at achieving greater equity, sustainability, and competitiveness, underpinned by strong economic management and governance. The evaluation recommended that the World Bank Group value benefits beyond individual interventions and make expected catalytic impact a major criterion in the design of its future strategy in Brazil.

Evaluations by Theme

Thematic evaluations can shed light on the overriding question of how development finance affects issues of inclusion, environmental protection, or governance, which form the core of the SDGs. Taken together, evidence emerging from OBE of individual projects or sectors can feed into the major themes that drive sustainable development in the aggregate. MDBs have done thematic evaluation on a number of themes: the environment and climate change (see, e.g., IDEV 2018; OVE 2014), gender and diversity (e.g., IEO of Global Environment Facility 2017; OVE 2018), and urban development (Cities Alliance 2017).

Consider a case where distributional impact is a concern and three projects—water access improvement, road construction, and deforestation activity control—were financed by an MDB. The water project may benefit the poor much more than the non-poor. Roads, on the other hand, may not connect the remote areas where the poor mostly reside, thus limiting the benefits that the poor can gain from them. It is also possible that controlling deforestation activity may on balance help the poor who depend on the sustainability of forest resources for their livelihood. Thus, different projects and investments in different sectors may affect the poor simultaneously but in different ways. Overall distributional impact can thus be assessed through a thematic evaluation.

Together, the various strands of OBE can be brought to bear on how much inclusion is taking place in the process of economic growth. This holds true also for sustainability and governance. Sustainability concerns projects that directly deal with environmental protection as well as projects in other sectors which might have an environmental impact. An evaluation of one project, or of a specific sector, is not capable of assessing the impact on environmental protection. Likewise, governance is related to projects in various sectors.

Inclusion

An ADB thematic evaluation on inclusive growth demonstrates the value thematic evaluations add to sectoral or country evaluations. ADB considers inclusive growth as a means to achieve poverty reduction, which was ADB's objective in its Strategy 2020. This thematic evaluation defined inclusive growth as growth with social equity, that is, a growth process in which all segments of the population can participate in and benefit from, particularly the poor (IED 2014b).

The evaluation highlighted inequality in terms of income as well as access to opportunities as a serious issue in the region. Although the region's gross domestic product (GDP) increased by 9 percent annually in the 1990s and by 8.2 percent in the 2000s, income inequality increased by about 1 percent annually in these two decades. Fast economic growth did not result in adequate access to opportunities either. Access to health care, water and sanitation services, and electricity varies widely across countries in the region and shows substantial gaps in many, for example, Bangladesh, Bhutan, Cambodia, Laos, Pakistan, and Vietnam.

The evaluation examined the promotion of inclusive growth through ADB's development finance between 2000 and 2012, which totaled US$137 billion (IED 2014a, b). It looked at the three pillars identified in ADB's strategy: (1) high sustainable growth to create and expand economic opportunities; (2) broader access to these opportunities to ensure that members of society can participate in and benefit from growth; and (3) safety nets to prevent extreme deprivation. As ADB's inclusive growth framework was broad, most of its projects were categorized under the three pillars, and ADB's support was heavily skewed toward pillar 1 (Fig. 4.3).

Given the rising inequality in the region, the evaluation suggested that the skewed focus on pillar 1 be reconsidered. The evaluation also emphasized the importance of project design and implementation in order to make projects inclusive. Considering urban-rural disparities are a major facet of unequal income distribution in the region, targeting rural areas, where the majority of the poor are found, could be an efficient way to accord special attention to benefiting lower-income groups. However, only 14.1 percent of infrastructure interventions targeted rural areas in the

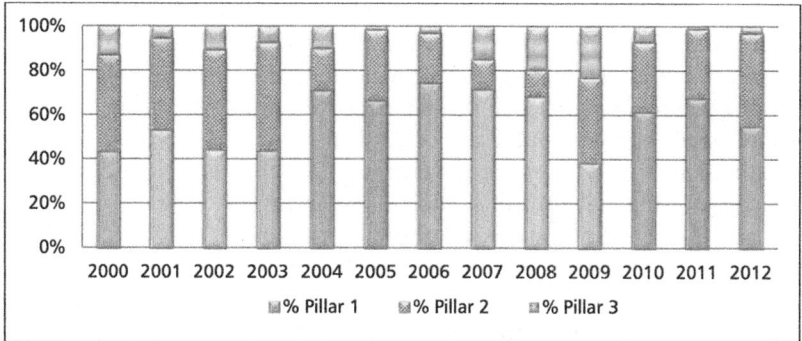

Pillar 1 = sustained and high growth; Pillar 2 = broader access to opportunities; Pillar 3 = strengthening safety nets.

Fig. 4.3 Share of yearly project amounts supporting inclusive growth by pillar, 2000–2012. (Source: IED 2014b)

13-year period. The study also suggested that the impact of infrastructure investments on inclusive growth be scaled up, for example, by linking rural infrastructures to schools, health centers, markets, and other services and opportunities.

The study assessed ADB's support toward the inclusive growth agenda at the country level in six countries. It found that the opportunities for and obstacles to inclusive growth vary by country, and it proposed that ADB's support toward inclusive growth should take into account the particular needs of each country and be designed in a way that maximizes the benefit to those who are poorer and with less opportunities.

Sustainability

The World Bank Group conducted a thematic evaluation of its support toward forest resource management for sustainable development in 2013 (IEG 2013). Matters of forest strategy would correspond to what was labeled in 2015 as SDG 15 on Life on Land. When the World Bank Group's 1991 Forest Strategy was implemented, international concern was directed toward protecting tropical forests. The Strategy, therefore, focused on protecting tropical moist forests by not financing commercial logging in primary tropical moist forests. However, a review by the IFC concluded that this did not contribute to combating global loss of forest (IFC 2000).

The review argued that this was because (a) the low stumpage paid for government-owned forests was lower than the real cost of sustainably managing the forests and did not provide adequate financial incentive for private operators to engage in sustainable reforestation responsibilities, (b) private operators were not given control over the forests on which their operations rely, and (c) many governments around the world desired to retain ownership and control of their forestlands (IFC 2000).

The 2002 Forest Strategy incorporated the findings and proposals of the review and reshaped its strategies based on three pillars: (1) protecting vital local and global environmental services and values, (2) harnessing the potential of forests to reduce poverty, and (3) integrating forests into sustainable economic development.

The changes made to the 2002 Forest Strategy are summarized in Table 4.9.

The changes highlight a geographical shift as well as the emphasis on sustainability. While the 1991 Strategy focused on tropical moist forests, the 2002 Strategy included all forest types. The 2002 Strategy also aimed to improve local livelihoods in Sub-Saharan Africa by protecting vast dryland forests and woodland areas, as the resources in these forests had

Table 4.9 Differences between the 1991 and 2002 World Bank Group Forest Strategies

	1991 Forest Strategy	*2002 Forest Strategy*
Forest focus	Tropical moist forests	All forest types
Priority countries	Forest-rich countries	Forest-rich and forest-poor countries
Thematic focus	Forest protection Resource creation Biodiversity conservation	Harnessing the potential of forests to reduce poverty Integrating forests into sustainable economic development Protecting vital local and global forest environmental services and values
Safeguards	Logging ban in tropical moist forests	Protecting critical natural habitats Independent verification of sustainable forest management
Implementation	Internal cooperation No internal strategy No incentive structure	Internal strategy developed based on selective engagement with partners

Source: World Bank, Progress Report on Implementation of the Revised Forest Strategy and Policy, Environmentally and Socially Sustainable Development Forest Team. August 2004; IEG (2013)

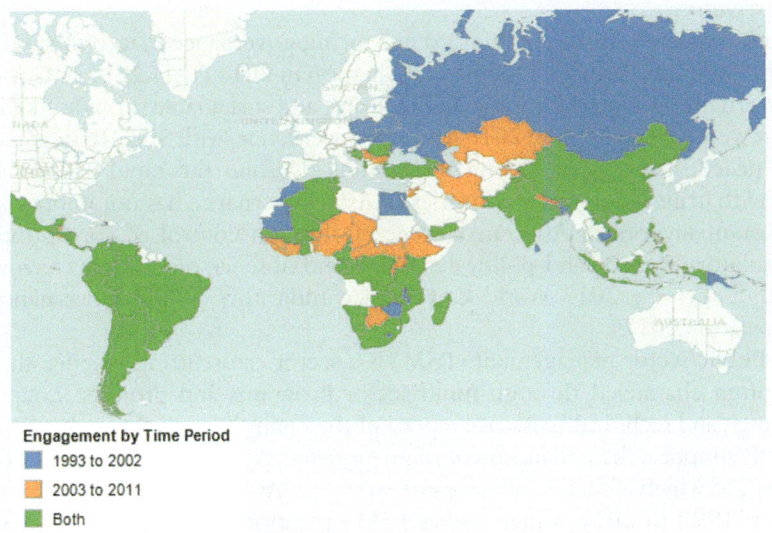

Fig. 4.4 World Bank forest activities before and after the 2002 Forest Strategy by country. (Source: IEG 2013)

commercial potential. As a consequence, there was a shift in emphasis toward Sub-Saharan Africa (Fig. 4.4).

IEG (2013) evaluated forest sector-related projects, supported by the World Bank Group member countries and private sectors between 2002 and 2011, based on whether they balanced competing demands on forest resources and at the same time were sustainably managed. It included 289 projects approved by the World Bank in 75 countries. The evaluation was conducted based on a review of the projects and portfolio as well as extensive interviews with stakeholders and site visits.

The evaluation concluded that the World Bank Group's forest interventions contributed positively to environmental outcomes. However, poverty reduction, for the most part, was not satisfactorily addressed. Among the interventions evaluated, the study found participatory forest management to have delivered livelihood enhancing benefits as well as positive environmental outcomes by linking forest products to markets. The study highlighted that participatory forest management should be promoted further. However, to make this intervention effective and sustainable, authority needed to be devolved to communities, and regulations needed to be improved to integrate small-scale informal forestry activities.

Governance

ADB evaluated its support toward enhancing governance in its public sector operations through a thematic evaluation in 2014 (IED 2014a). Good governance is crucial for achieving inclusive and sustainable growth. SDGs that were agreed upon in 2015 featured governance, with SDG 16 devoted to peace, justice, and strong institutions. Despite the rapid economic growth achieved in many countries in Asia, governance has not improved substantially. South Asia ranks especially low on control of corruption, regulatory quality, and political stability and absence of violence/terrorism, as per the 2017 World Governance Indicators (World Governance Indicators 2017).

Public sector management (PSM) is a sector-crosscutting practice that is often channeled through multi-sector programs and projects. Loans, grants, and technical assistance represent the main channels through which ADB supports the enhancement of governance. ADB financed US$11 billion (of which US$2.3 billion came from the Asian Development Fund) from 1999 to 2011, which makes PSM operations ADB's fourth largest sector program during this period. A review of the success rate in 1990–1999 and 2000–2010 showed an improvement in the latter period. However, compared to other sectors, performance of PSM financing is still weak.

The analysis of failed PSM projects underscored three common factors: lack of institutional capacity and/or resources in government to undertake projects or reforms, overly ambitious or complex designs, and weak government ownership and commitment (IED 2014a). The thematic evaluation recommended more rigorous diagnostics to be conducted at the project design stage, which required sufficient understanding of institutional capacity and political commitment and incentives.

Thematic evaluations thus help MDBs identify successes as well as gaps in their development financing. Through these evaluations, MDBs can enhance their impact on SDGs that form the core of their mission.

Conclusion

In Chap. 2 we discussed the IE tools that enable evaluators to assess the extent and nature of the difference a development project makes. CBA, as discussed in Chap. 3, is a complementary approach, providing indications of the net gains derived from a project. OBE can draw on both these tools in assessing the degree to which the goals of a project are being met. Its primary emphasis is not so much the causal effect of a project, but more on the contribution the project makes in meeting stated objectives of sustainable development. Even with a counterfactual analysis built in, the results are associations, as they do not adequately control for the simultaneous effects of other projects and interventions.

OBE thus provides an overarching understanding of the effectiveness of development projects. Both external agencies and country governments can get an overview of the objectives and how they are being met under a set of common criteria, albeit with differences across different financing agencies. Evaluations of projects can be aggregated into results for sectors, themes, or countries. Such evaluations can help inform how much of the original or revised objectives are being achieved and what might be done to improve performance. Thus, OBE is a good check on accountability with respect to sustainable development that also provides lessons for future development financing.

Bibliography

Cities Alliance. 2017. *Programme Evaluation of Cities Alliance Country Programmes: Ghana, Uganda and Vietnam.*

EBRD (European Bank for Reconstruction and Development). 2013. Evaluation Policy. *Policy Document.* London: European Bank for Reconstruction and Development.

Gertler, Paul J., Sebastian Martinez, Patrick Preman, Laura B. Rawlings, and Christel M. J. Vermeersch. 2016. *Impact Evaluation in Practice.* Washington DC: World Bank.

Guardian. 2013. "Disaster Resilience: The Private Sector Has a Vital Role to Play." https://www.theguardian.com/sustainable-business/disaster-resilience-private-sector-role.

IDEV (Independent Development Evaluation). 2015. Evaluation of the AfDB Assistance in the Energy Sector. Abidjan, Côte d'Ivoire: African Development Bank.

IDEV (Independent Development Evaluation). 2018. Evaluation of the Congo Basin Forest Fund: A Thematic Evaluation. Abidjan, Côte d'Ivoire: African Development Bank.

IED (Independent Evaluation Department). 2014a. ADB Support for Enhancing Governance in its Public Sector Operations. Manila: ADB.

IED (Independent Evaluation Department). 2014b. ADB's Support for Inclusive Growth. Manila: ADB.

IED (Independent Evaluation Department). 2016. Guidelines for the Economic Analysis of Projects. Manila: ADB.

IED (Independent Evaluation Department). 2017. India, 2007–2015. Manila: ADB.
IEG (Independent Evaluation Group). 2006. From Schooling Access to Learning Outcomes: An Unfinished Agenda. Washington DC: World Bank.
IEG (Independent Evaluation Group). 2007. A Decade of Action in Transport: An Evaluation of World Bank Assistance to the Transport Sector, 1995–2005. Washington DC: World Bank.
IEG (Independent Evaluation Group). 2009. Knowledge for Private Sector Development. Washington DC: World Bank.
IEG (Independent Evaluation Group). 2010. Priorities in Meeting the MDGs: Lessons from Evaluation. Washington DC: World Bank.
IEG (Independent Evaluation Group). 2013. Managing Forest Resources for Sustainable Development: An Evaluation of World Bank Group Experience Washington DC: World Bank.
IEG (Independent Evaluation Group). 2015. Project Performance Assessment Report: Brazil: First Programmatic Development Policy Loan for Sustainable Environmental Management. Washington DC: World Bank.
IEG (Independent Evaluation Group). 2017a. IEG Methodology. Washington DC: World Bank.
IEG (Independent Evaluation Group). 2017b. Second Community Water Supply and Sanitation Project. Washington DC: World Bank.
IEO (Independent Evaluation Office) of Global Environment Facility. 2017. Evaluation of Gender Mainstreaming in the GEF. Washington DC: Global Environment Facility.
IEO (Independent Evaluation Office) of IMF. 2016. The IMF and the Crises of Greece, Ireland and Portugal. Washington DC: IMF.
IEO (Independent Evaluation Office) of UNDP. 2018. Independent Country Programme Evaluation: Bhutan. New York: UNDP.
IFC (International Financial Corporation). 2000. OEG Review – Implementation of the 1991 Forest Strategy in IFC's Projects. Washington DC: IFC.
IFC (International Financial Corporation). 2014. End of Program: Africa Schools. IFC.
IOE (Independent Office of Evaluation). 2017. Participatory Natural Resource Management Programme. International Fund for Agricultural Development.
Kanbur, Ravi, and Andy Sumner. 2012. "Poor Countries or Poor People? Development Assistance and the New Geography of Global Poverty." *Journal of International Development* 24 (6):686–695.
MDBs (Multilateral Development Banks). 2018. Multilateral Development Banks' Harmonized Framework for Additionality in Private Sector Operations. Multilateral Development Banks.
Mohler, G. O., M. B. Short, S. Malinowski, M. Johnson, G. E. Tita, A. L. Bertozzi, and P. J. Brantingham. 2015. "Randomized Controlled Field Trials of Predictive Policing." *Journal of the American Statistical Association* 110 (512):1399–1411.

OECD-DAC (Organisation for Economic Co-operation and Development–Development Assistance Committee). 1991. DAC Principles for Evaluation of Development Assistance. Paris: Development Assistance Committee
OED (Operations Evaluation Division). 2018. Evaluation of the European Fund for Strategic Investments. European Investment Bank.
OVE (Office of Evaluation and Oversight). 2014. Climate Change at the IDB: Building Resilience and Reducing Emissions. Inter-American Development Bank.
OVE (Office of Evaluation and Oversight). 2015a. Country Program Evaluation: Brazil 2011–14. Inter-American Development Bank.
OVE (Office of Evaluation and Oversight). 2015b. Review of the Bank's Support to Agriculture, 2002–14. Inter-American Development Bank.
OVE (Office of Evaluation and Oversight). 2018. Evaluation of the Bank's Support for Gender and Diversity. Inter-American Development Bank.
Petri, Peter, and Vinod Thomas. 2013. Development Imperatives for the Asian Century. *ADB Economics Working Paper Series no. 360*. Manila: ADB.
World Bank. 2009. Global Monitoring Report 2009: A Development Emergency. Washington DC: World Bank.
Worldwide Governance Indicators. 2017.

Open Access This chapter is licensed under the terms of the Creative Commons Attribution 4.0 International License (http://creativecommons.org/licenses/by/4.0/), which permits use, sharing, adaptation, distribution and reproduction in any medium or format, as long as you give appropriate credit to the original author(s) and the source, provide a link to the Creative Commons licence and indicate if changes were made.

The images or other third party material in this chapter are included in the chapter's Creative Commons licence, unless indicated otherwise in a credit line to the material. If material is not included in the chapter's Creative Commons licence and your intended use is not permitted by statutory regulation or exceeds the permitted use, you will need to obtain permission directly from the copyright holder.

CHAPTER 5

Conclusion and Future Directions

Abstract This concluding chapter summarizes our discussion and sets out three aspects that can help make economic evaluation of sustainable development stronger—broadening our understanding of the direct and indirect impacts, recognizing the global dimension of sustainable development priorities, and getting innovative with data to make evaluations current and relevant.

Keywords Innovation • Development priorities • Big data • Global public goods • Indirect impacts

> *The whole of the international community has to shoulder a responsibility to bring about a sustainable development.*
> Angela Merkel

We started the discussion on evaluation by highlighting the overarching theme of sustainable development that comprises growth with inclusion, environmental stewardship, and good governance. These themes are present in the development plans and discussions of countries, and, in varying measure, they are essential ingredients of societal visions. Sustainable Development Goals capture this desired direction with targets for 17 attributes to be achieved by countries by year 2030.

EVALUATION OF SUSTAINABILITY

This book makes the case for a stronger pursuit of the Sustainable Development Goals (SDGs) by according economic evaluations their rightful role in development work. The degree to which these goals are being met often falls short of expectations. However, there are significant welfare gains from policies that more effectively help achieve them. In each of the chapters dealing with impact evaluation (IE), cost-benefit analysis (CBA), and objectives-based evaluation (OBE), we have seen illustrations of how the value of interventions might be enhanced.

One way to put greater energy and drive into achieving progress toward the SDGs is to have better and timely assessments of sustainability—much as the experience with the previous Millennium Development Goals (MDGs) showed (IED 2013). But our ability to evaluate how well SDGs are being achieved is patchy. Ways and means for assessing progress need to be pursued and continuously improved, as examples in Chaps. 2, 3, and 4 illustrate. There is room for developing capacity for undertaking such evaluations and for adequately funding the efforts across countries.

A key factor in ensuring sustainable development would be the political support across countries and at various levels of governance. And one way to garner political support is to embed the results of evaluation much more frontally in the policy agenda of countries and global financial institutions. Timely release of the findings and their transparent application in decision-making help, again, as the experience with MDGs demonstrated.

Bringing evaluation to bear on the goals of sustainable development has been one overriding objective of this book. While Chaps. 2, 3, and 4 did not evaluate the achievement of the SDG targets per se, the different approaches to evaluation picked up the goals of sustainable development, albeit with gaps. As indicated in Chap. 1, economic growth forms an integral part of the evaluation as the very matrix of measuring value addition, benefits, and costs, or welfare gains is often the change in output or GDP. The challenge is how to put inclusion, sustainability, and governance under an evaluative lens, side-by-side with economic growth.

While this broadens the scope of assessments, focus and rigor should not be compromised. To be effective, it is essential for the work to be well-focused, well-defined, and rigorous. The broader focus should allow the evaluation to triage actions and options toward sustainable development.

The interplay of evaluation and economics helps in making the decision of the choice of topics and the scope of the work (see Van den Berg et al.

2018). Presenting stronger ties between economics and evaluation has been a second objective of the book. We have seen how assessments of sustainable development can be done carefully and credibly by applying tried-and-true tools of IE, CBA, and OBE. Such work can span from the micro and project-level assessments to the macro and aggregative assessments. But in either case, the application of economic analysis and evidence can bolster evaluations.

Quantifying costs and benefits of interventions to reflect distributional considerations can be aided by economic analysis of growth impacts on changes in income distribution (Dabla-Norris et al. 2015; Ostry et al. 2014). The assessment of global spillovers can be assisted by featuring health and climate change externalities (Sommer 2016; IMF 2015). Objectivity of information can be enhanced by the complementary data often culled directly from sources, for example, weather data connecting knowledge on weather patterns, high-risk areas, and people at risk (Emmanouil and Nikolaos 2015).

In this final chapter, we go further to see how the frameworks in Chaps. 2, 3, and 4 can be extended to get more mileage on sustainability. We set out three aspects that can help make economic evaluation of sustainable development stronger. First, there is much to be gained by looking for and into the important linkages—both direct and indirect—that contribute to outcomes. For example, indirect and non-income aspects (Dennig 2017) are important in considerations of inclusive growth. Second, sustainable development challenges have a local component and a global part (Everett et al. 2010). Often the local effects are mostly intended while the global carry an unintended component. Third, evaluators can innovate the data being used: with the availability of big data from the internet and social media, a huge window of opportunity has been thrown open (Faghmous and Kumar 2014).

Direct and Indirect Impacts

Broadening the field for evaluation helps to identify linkages that trigger important positive or negative impacts—across areas of concern or over time. Human well-being, which underlies inclusive growth, is understood to be multidimensional, including aspects of not only income but also education, health, and life satisfaction. These attributes, which are seldom incorporated in assessments of growth, might have significant impacts. The same can be said for environmental protection, where increasing growth is not enough to generate sustainable outcomes and where lack of environmental care itself can stunt growth. The 17 goals under the SDG framework help put emphasis on these non-monetary attributes of well-being.

Well-Being and Inclusion

One limitation often found in evaluation studies is their sole focus on impacts of interventions that are immediately observable. Usually, economic evaluation primarily concentrates on direct effects on income or expenditure. However, going from outputs to outcomes and impacts (as shown in Fig. 1.2) requires evaluation of sustainable development to look beyond immediately observable outcomes and to broaden its lens to focus on outcomes and impacts. While income and expenditure are useful measurements in that they are objective and clear, they do not fully capture the essence of sustainable development as it pertains to well-being and inclusion.

Human well-being in the context of sustainable development incorporates human capital, subjective well-being, and equal opportunity among other things (Thomas et al. 2000; Sachs 2012). The evaluation questions and the goals against which sustainable development is evaluated need to incorporate these as well.

As an example, an evaluation of an education project should not only be about increasing access but also about augmenting human capital. The goals should include immediate outputs such as the construction of more schools and also longer-term outcomes such as building knowledge and skills. A look at previously excluded groups, in particular girls, is most important. While the SDGs in themselves are sensitive to these differences, evaluations are yet to catch up. Most targets under the goal on quality education (SDG 4) have a gender parity component.

Accounting for interactions and spillovers of policies and projects with subjective well-being might make evaluations more meaningful. As an example, a study on unreliable urban water supply in the Kathmandu Valley in Nepal examines impacts of household coping costs (including those for collecting, pumping, purchasing, storing, and treating water) on well-being, which captures both evaluative (life satisfaction) and hedonic (feeling and emotions) reactions (Chindarkar et al. 2018). Findings reveal that coping cost is positively correlated with life satisfaction.

This seemingly counterintuitive finding is explained by households' perception of coping costs as investments in household health and ability to be resilient. An insight for policy-makers from this evaluation is that under conditions of policy inaction, as has been the case with worsening urban water supply in Kathmandu, households need to spend time and money on coping with unreliable water supply to sustain their well-being and develop resilience. Thus, conclusions from evaluations that consider subjective well-being as an outcome could be different and insightful.

The scope for using well-being as an outcome is broad. Measures of subjective well-being can be used to assess impacts of environmental problems such as air pollution and climate change as well as incorporated in CBA (Chen et al. 2019; Li et al. 2014; Rehdanz and Maddison 2005). Studies have also examined the impact of rising inequality on subjective well-being (Graham and Felton 2006; Jiang et al. 2012).

Gaps in opportunities stem from differences in access to education, health, and other basic services. Yet when it comes to assessing the impact of growth and other policies on inequality, outcomes are often restricted to monetary measures such as mean log deviation, Gini coefficient, and Theil index. The limitation of these measures is that by focusing on income distributions they look only at the observed outcome of unequal opportunities and not at the unequal distribution of opportunities themselves that underlie individual advantage or disadvantage.

One proposition to evaluate equality of opportunity itself is to examine differences based on "circumstances," such as place of birth, gender, and parental characteristics, over which individuals have no influence (Roemer 1993). In recent years attempts have been made to develop indices that capture lack of opportunities. Important among these is the human opportunity index (HOI) developed by De Barros, Ferreira, Vega, and Chanduvi (2009).

The HOI focuses only on dependent children and measures inequality of opportunity in terms of access to education, health, sanitation, and other basic services. The rationale for this focus is that access to these basic services is exogenous to children and therefore constitutes a circumstance for them. However, studies evaluating the inequality inherent in policies such as those relating to school construction or provision of development finance have overlooked HOI to see how policies affect shifts in distribution of opportunities.

Environmental Protection

With increasing pressure on the use of natural resources and runaway climate change triggered by the build-up of emissions in the air, environmental stewardship needs to take center stage in evaluations. Underlying much of environmental care is the understanding that natural capital, along with physical and human capital, is an integral part of the framework on how economic growth is generated (Thomas et al. 2000). There are evaluations of individual aspects of environmental impacts, but assessments of how the environment impinges on overall sustainability are lacking.

In a framework of sustainable development, economic growth is generated by investments in physical and financial capital, human and social capital, and environmental and natural capital. Economic policies by and large have favored investment in physical and financial capital through various forms of subsidies. Human and social capital have received increasing investments over the decades, but evaluations need to pay more attention to the degree of underinvestment seen when taking into account the positive externalities being generated.

Environmental and natural capital are not generally invested in, rather there is much degradation and unsustainable use. There is room to evaluate how this gap affects growth and sustainable development. If nature is included as a capital asset in production activities, there is likely to be a concern over growth patterns that conflict with the achievement of sustainable economic development. It would be useful to assess how the accumulation of physical and human capital may not have compensated for the degradation of natural capital.

The broader evaluative framework would allow evaluators to make direct connections and assess spillovers and indirect impacts among investments in different forms of capital. For instance, greater provision of environmental services can have the direct and tangible benefits such as lesser air and water pollution, which in turn can generate broader gains for worker productivity and livelihood (Zivin and Neidell 2012).

Broader Goals in Asia

As an application of a broader framework, we might consider the developments in Asia. Economic growth remains the biggest driver of development aspirations, but the vital linkages of other attributes to growth are emerging.

The need for evaluation to factor in social inclusion and the environment come through prominently in the case of Asia. Income inequality has worsened over the last decade in China, India, and other countries that, taken together, account for 80 percent of the region's population. Developing Asia's Gini coefficient went from 0.39 in the mid-1990s to 0.46 in the late 2000s. Furthermore, developing Asia is the world's leading emitter of greenhouse gases, accounting for nearly 40 percent of global emissions, twice its share of global GDP. Air pollution is now at dangerously high levels in many Asian cities, notably New Delhi and Beijing, and environmental degradation is worsening across the region.

Incorporating and addressing gender inequality is a crucial dimension of inclusion. It is estimated that close to 100 million women are "missing" in Asia owing to gender-discriminatory practices (ADB 2012). Women in Asia are found to be worse off compared to men across various dimensions including health, access to education, asset ownership, and political inclusion (ADB 2012). Sensitizing evaluations to gender equality by explicitly incorporating gender-sensitive indicators would be a huge step forward. Gender-sensitive indicators such as maternal health, time use, and distributive impacts can be explicitly incorporated in IE, CBA, and OBE. The SDG framework on gender equality (SDG 5) and other goals where gender parity is considered can help shed light on gender-development issues.

Evaluation would want to take on board research results showing the deleterious effect of poor governance on growth (Kaufmann et al. 2010). Asia presents a mixed picture in global measures of good governance. For example, Southeast Asian countries in general fare poorly in their control of corruption in governance surveys, and this can affect growth drivers, including foreign investment and credit ratings. In East Asia, the gaps are wide for voice and accountability—an indicator which captures perceptions of the extent to which citizens can participate in policy-making processes and the accountability of governments. South Asia ranks low in political stability.

An example of how policy and strategy can guide evaluations toward achieving these broader goals is Asian Development Bank's (ADB) new 2030 strategy. This strategy stresses sustainable development beyond economic growth in terms of greater inclusion, resilience, and well-being (ADB 2018). The approaches to bring about such progress are to be "integrated and multi-disciplinary" in order to address the complex problems of "inequality, climate change and urbanization which cut across several sectors." Development financing under this strategy is explicitly aimed at achieving well-being, inclusion, and climate mitigation and adaptation, and incorporates these as evaluation goals. The challenge is how to make these directions operational.

Local and Global Public Goods

Development priorities and challenges are increasingly taking on global dimensions. Local issues, like deforestation or slash-and-burn practices in one country, can affect neighboring countries (Thomas 2018). A case in point is Indonesia, where each year slash-and-burn agriculture causes massive emissions that hurt the health of populations not only within Indonesia but also in neighboring Malaysia, Singapore, and beyond. In this case as well as the case of massive air pollution and smog in Asia's megacities, the local effects spillover to regional and even worldwide scales, aggravating global warming. Another example is the global financial crisis, which originated in a few centers in the developed world, but its social effects in terms of increased inequality and poverty rippled across the world.

Global efforts are called for as scientists make clear vast biodiversity losses and rapid climate change across the globe. The world has lost 60 percent of the animal life on the planet since 1970 (WWF 2018), and global warming is estimated to reach a critical level by as early as 2030 (IPCC 2018). Evaluations must move from a growth-only focus and pay considerably more attention to these urgent issues.

Governance too has global dimensions. Studies show that more open trade and globalization have brought net gains to countries in many instances (e.g., IMF Staff 2001; Dabla-Norris and Duval 2016). But there are also losers, and at times their interests, true or perceived, can dominate. The world has witnessed the United States government reneging on global agreements on emissions and international trade. This highlights the role of opposing interests and the fact that even where the aggregate gains are positive, the interests of particular groups that might lose become decisive inputs into policy.

Special efforts are needed to assess and share the findings about the gains for common goods from collective actions, especially where global public goods (GPGs) are involved. Important themes for future evaluations are the effectiveness of global funding mechanisms such as climate change funds, multilateral agreements such as regional economic partnerships and global climate agreements, and bilateral agreements such as transboundary water conventions.

Evaluating issues and policies pertaining to GPGs is complex, which probably explains why the evaluation techniques discussed in previous chapters—IE, CBA, and OBE—do not systematically incorporate GPGs. Complexities also pertain to funding evaluations of GPG interventions and the institutional setup required to conduct these evaluations.

Kanbur (2017) argues that by its very nature, the benefits of addressing transboundary issues are also transboundary. Since benefits accrue beyond individual countries, incentives, such as grants, are needed to motivate countries to collaborate on GPGs. By extension, financing evaluations of GPG interventions would also require setting up grants and collective deliberation on performance indicators. While each country should have its own platform for implementing actions addressing GPG issues, evaluation institutions and mechanisms are required at the global level. A way forward is to build in independent evaluations into the global mechanisms to assess transboundary benefits.

Attention would need to be given to spillover effects when evaluating GPG interventions. We have discussed how spillovers can be incorporated into IE and CBA in Chaps. 2 and 3. The same ideas can be extended to transboundary spillovers. To design an experimental or quasi-experimental IE of a GPG intervention, evaluators would first need to have good knowledge (based on theory or prior evidence) of why and how spillover effects occur.

The treatment and control groups in IE can then be identified in the relevant socioeconomic unit (group of regions or group of countries) within which the spillovers occur, and treatment effects can be adjusted to avoid biased estimates (Angelucci and Di Maro 2015). Similarly, in CBA the relevant socioeconomic unit of analysis will need to be identified and marginal social cost be adjusted based on whether the transboundary spillover is positive or negative, and consequently net social benefit would be altered.

A further complexity in evaluating GPG interventions is reliable data. Little is known about the spending by countries on GPGs. Some attempts are being made to estimate these outlays such as those by Birdsall and Diofasi (2015). However, this is just a start and better reporting practices and, more fundamentally, an agreement on what should count as spending on GPGs is needed. For instance, spending on HIV/AIDS prevention in Africa by the United States can be thought of primarily as financing for treating and preventing the disease within-country boundaries. However, given the large out-migration from Africa, HIV/AIDS prevention also has GPG characteristics.

A related challenge is that of suitable methodological tools. While entirely new tools are probably not required, what is required is pliability of the tools reviewed in this book in evaluating GPG interventions. For instance, an IE of development finance on global HIV/AIDS treatment and prevention should estimate and disaggregate the average treatment effect by within- and between-country treatment effects. More work is needed on refining the econometric tools for evaluating transboundary effects. A CBA of HIV/AIDS financing should account for the fact that net social benefits are not restricted to affected countries but also have implications for countries where people from affected countries migrate to. The same central criteria for OBE can be used but with specific attention given to transboundary effectiveness, efficiency, development impact, and sustainability.

Van den Berg (2011) cautions that evaluation of GPGs can show a "micro-macro paradox." This term refers to a situation where local (or within-country) interventions might be successful, yet when assessed at the global level, the interventions do not translate into desirable outcomes. For example, individual countries might achieve emissions reductions through carbon taxes. However, at the global level there might be no observed change or even an increase in emissions if industrialized countries shift pollution-generating activities to less developed countries.

Similarly, an evaluation of development finance might find that it does achieve SDGs in individual countries; however, the global impact of development finance might be limited. In this case, the micro-macro paradox can partly be explained as a consequence of insufficient public funding available to meet global public costs such as for climate-induced disasters or forced migration. These paradoxes offer lessons on interventions that have different local, regional, or global effects.

Small and Big Data

Sound evaluations are invariably predicated on sound data. For the most part, these data have come from household, national, and international survey and estimations, made available to researchers in published and unpublished forms. Gaps are serious, particularly on many aspects of sustainable development. Greater attention to evaluation of sustainable development should motivate more investments in generating and sharing the underlying data.

The explosion of digital technology and the expanding amount of data now hold promise in enabling their use in research and evaluation. The application of big data is quickly expanding in business, government, and civil society. For example, various agencies of the United States' government at the central and state levels are mining and analyzing data to mitigate fraud, enforce law, and monitor usage of resources. An example is the Department of Health and Human Services, which implemented a fraud prevention system identifying millions of dollars in improper payments to health-care providers (see, e.g., US GAO 2017).

Other examples include Australia, the Nordic countries, and the United Kingdom, where governments track citizens through the course of their lives. The data they collect contain information on birth outcomes, education outcomes, and health outcomes, which are then linked with socioeconomic information, creating a rich database that is ideal for policy evaluation. Large-scale administrative data are also being sourced from utility bills, public-transport smart cards, banking and credit card transactions, satellite images, and so on.

Application of big data to evaluations is mostly confined to identifying correlations and predicting trends (UN Global Pulse 2012). While this is quite different from counterfactual IE, correlations and trends generated from large volumes of data can still be useful as they may closely represent the population. Correlations can be used to identify systematic patterns and repeated behaviors, consequently unveiling stylized facts about inclusive growth, sustainability, and governance.

For instance, predictive analysis can help identify students at risk of dropping out. Monitoring student retention rates will make way for enhancing student academic performance and therefore satisfaction among students, teachers, and school administration. Data gathered on individual students' learning styles can also assist teachers, who can adjust their teaching styles according to students' needs.

The World Health Organization declared the Zika virus a global health emergency in 2016 and forecasted the spread of the virus. While there were no reliable tests and vaccines for the virus at the outset, utilizing data-driven infrastructure helped to identify trends and analyze clinical-test results, shortening the search for a cure. Health systems are using big-data technologies like Apache Hadoop to take real-time streams of data from monitors, machines, and wearables and combining it with electronic health records (Juric et al. 2017). Big-data technologies make it possible to apply intelligence to multiple electronic data feeds of clinical tests as they stream in.

As mayors struggle to make cities financially viable and sustainable, big data can be used to create "smart" cities. Every city has its own intricacies, and therefore no master design exists, but every smart city presents an opportunity for big data to govern public policies. For example, the city of Boston uses the crowdsourcing app Street Bump to collect data from citizens' smartphones to allocate maintenance and repair crews, resulting in vast savings (Zie 2015). In San Francisco, smart meters provide digital reads of water flow to track citizens' water usage, also producing sizable savings.

The use of big data is proving to be a valuable tool in disaster management. Advances in ground-based networks of radars as well as in satellite data are key to nearly continuous observation of global weather. Japan's Meteorological Agency recently updated its Evaluation Alert System with much more detailed data to support evacuations, mapping the intensity of weather-related hazards and people with special needs. In Turkey, a new National Emergency Management Information System and an Uninterrupted and Secure Communication System Project link authorities during emergencies. Australia's Emergency Alert enables territories to issue warnings through landline and mobile telephones linked to high-risk properties, working across telecommunication carrier networks.

Technologies that link sensor networks, large-scale data analysis, and communications systems can provide decision-makers with timely information to guide responses. Siemens implemented a levee monitoring system in the Netherlands using sensors to monitor water pressure, temperature, and shifting weather patterns to identify areas that are at risk of being breached and trigger alarms (Guardian 2013). IBM provided a digital command center that integrates real-time information on storm conditions, emergency-response assets, and areas at risk (Guardian 2013).

Pertaining to governance, law enforcement is another area benefiting from big data. The implementation of predictive policing is relatively new, and it is currently being tested and deployed in several cities across the United States. The method uses data from type, place, and time of previously committed crimes in order to assign probabilities of future crime incidents. In some places, there is evidence of a decline in crime as a result (Mohler et al. 2015).

Developing countries are benefiting from such real-time evaluations to track their progress toward achieving the SDGs. Case in point, a laboratory in Rwanda uses electronic sensors to assess the use of water filters and cook stoves. The UN Global Pulse runs a number of projects that use data from social media to monitor social and environmental issues. One project analyzes conversations on social media to understand public perceptions on sanitation, providing a baseline for change in public discourse on sanitation.

Similarly, other Global Pulse projects use Twitter to measure global engagement on climate change. Food security issues are also being assessed, as in Indonesia, where the correlation between actual food price fluctuations and perceptions about food inflation on Twitter were tracked (UN Global Pulse 2011). Comparison of the trends of actual food price fluctuations and tweets on food inflation shows that public perception about food inflation on social media closely tracked actual prices.

Evaluators have attempted to use big data for causal analysis by applying experimental and quasi-experimental tools to a large pool of observations. Ibarra, McKenzie, and Ortega (2017) use high-frequency financial data on over one hundred thousand credit card clients in Mexico to evaluate the impact of financial education on credit card usage and bill payment behavior. They find that while financial education increases the probability of paying bills on time and paying more than the minimum payment due, it does not reduce spending.

Big data can complement traditional IE, survey data, and official statistics by adding up-to-date information to provide a fuller picture of evaluations. However, there are several things to bear in mind when using big data for evaluation. While these data open up avenues for innovative evaluations, evaluators must exercise caution, particularly when it comes to privacy and personal data protection. When accessing and using these data, evaluators must be aware of country laws pertaining to data protection and undergo the required review process to get approval for conducting their IEs. Also, big data might contain inherent selection bias in countries where internet and smartphone penetration is low. In these cases, big data will only reflect the behaviors and opinions of those with access to technology. And finally, big data cannot fully replace, but only complement, rigorous evaluations.

Conclusion

In Chaps. 2, 3, and 4, we have seen applications of IE, CBA, and OBE in assessing performance and providing lessons for improvements. As countries, in varying degrees, now embrace the SDGs, it is crucial for evaluations to keep its eye on monitoring and tracking sustainable development. Aiming for sustainable development also helps to achieve a greater integration of work across sectoral and thematic boundaries, such as infrastructure and the environment or education and labor markets.

Our examples also suggest that there are gains in taking advantage of the interplay between evaluation and economics. For example, evaluations of economic growth and income distribution are much more impactful when they bring together findings from an economic theory of change and evaluation. The quality of the data, the rigor of analysis, and the timeliness of the findings all decide how useful the work is and how influential it is in shaping decisions and policy.

There are fruitful avenues for evaluation to capture social inclusion, environmental care, and good governance, in addition to economic growth. Incorporating regional and global effects beyond the local level is becoming increasingly essential. These effects are immensely important, for example, in income inequality and climate change. Innovative data may lend themselves to addressing these broader questions and help deliver better results.

Employing a broader development lens in individual evaluations has been a challenge. Broadening the agenda, even when it makes eminent sense, introduces complexities and difficulties, not least of them being the limits placed by the availability of methods and data. It is important that in broadening the scope of work, one does not lose sight of the priorities in terms of the outcomes and of the needed selectivity in terms of the most important linkages that matter.

In the end, the quality of the evaluation work determines the usefulness of the findings for policy-making. Broadening the evaluative lens strengthens the relevance of findings and improves the chances of capturing crucial indirect and unintended effects of interventions. At the same time, broadening of the field needs to ensure rigor, comparability, and some degree of replicability of the findings. In this respect, aligning evaluations with a commonly agreed set of goals and aspirations such as the SDGs, tracking progress, and drawing on lessons of experience will help.

Bibliography

ADB (Asian Development Bank). 2012. Gender Equality and Discrimination in Asia and the Pacific: 12 Things to Know. Manila: ADB.

ADB (Asian Development Bank). 2018. Strategy 2030: Achieving a Prosperous, Inclusive, Resilient, and Sustainable Asia and the Pacific. Manila: ADB.
Angelucci, Manuela, and Vincenzo Di Maro. 2015. "Program Evaluation and Spillover Effects." *Journal of Development Effectiveness* 8 (1):22–43.
Birdsall, Nancy, and Anna Diofasi. 2015. Global Public Goods for Development: How Much and What For. Washington DC: Center for Global Development.
Chen, Shuai, Ping Qin, Jie-Sheng Tan Soo, and Chu Wei. 2019. "Recency and Projection Biases in Air Quality Valuation by Chinese Residents." *Science of the Total Environment* 648:618–630.
Chindarkar, Namrata. 2015. "Why Public Policy Needs to Take a Broader View on Well-Being." In *Governing Asia: Reflections on a Research Journey*, edited by Lee Kuan Yew School of Public Policy. Singapore: World Scientific.
Chindarkar, Namrata, Yvonne Jie Chen, and Yogendra Gurung. 2018. "Subjective Well-Being Effects of Coping Cost: Evidence from Household Water Supply in Kathmandu Valley, Nepal." *Journal of Happiness Studies*. https://doi.org/10.1007/s10902-018-0060-6.
Clark, Andrew E., Paul Frijters, and Michael A. Shields. 2008. "Relative income, happiness, and utility: An explanation for the Easterlin paradox and other puzzles." *Journal of Economic Literature* 46 (1):95–144.
Dabla-Norris, Era, and Romain Duval. 2016. "How Lowering Trade Barriers Can Revive Global Productivity and Growth." https://blogs.imf.org/2016/06/20/how-lowering-trade-barriers-can-revive-global-productivity-and-growth/.
Dabla-Norris, Era, Kalpana Kochhar, Nujin Suphaphiphat, Frantisek Ricka, and Evridiki Tsounta. 2015. Causes and Consequences of Income Inequality: A Global Perspective. *IMF Staff Discussion Note*: Washington, DC.
Dasgupta, Partha. 2017. "The Idea of Sustainable Development." *Sustainability Science* 2 (1):5–11.
De Barros, Ricardo Paes, Francisco H. G. Ferreira, Jose R. Molinas Vega, and Jaime Saavedra Chanduvi. 2009. Measuring Inequality of Opportunity in Latin America and the Caribbean. Washington DC: Palgrave Macmillan and The World Bank.
Dennig, Francis. 2017. "Climate Change and the Re-Evaluation of Cost-Benefit Analysis." *Climatic Change* 1–12.
Emmanouil, Dontas, and Doukas Nikolaos. 2015. Big Data Analytics in Prevention, Preparedness, Response and Recovery in Crisis and Disaster Management. *Recent Advances in Computer Science (Proceedings of the 19th International Conference on Computers)*, edited by Xiaodong Zhuang. Zakynthos Island, Greece.
Everett, Tim, Mallika Ishwaran, Gian Paolo Ansaloni, and Alex Rubin. 2010. Economic Growth and the Environment. *MPRA Paper no. 23585*: Department for Environment Food and Rural Affairs.

Faghmous, James H., and Vipin Kumar. 2014. "A Big Data Guide to Understanding Climate Change: The Case for Theory-Guided Data Science." *Big Data* 2 (3):155–163.
Graham, Carol, and Andrew Felton. 2006. "Inequality and Happiness: Insights from Latin America." *Journal of Economic Inequality* 4 (11):107–122.
Guardian. 2013. "Disaster Resilience: The Private Sector has a Vital Role to Play." https://www.theguardian.com/sustainable-business/disaster-resilience-private-sector-role.
Gurung, Yogendra, Jane Zhao, K. C. Bal Kumar, Wu Xun, Bhim Suwal, and Dale Whittington. 2017. "The Costs of Delay in Infrastructure Investments: A Comparison of 2001 and 2014 Household Water Supply Coping Costs in the Kathmandu Valley, Nepal." *Water Resources Research* 53 (8):7078–7102.
Harberger, Arnold. 1978. "On the Use of Distributional Weights in Social Cost-Benefit Analysis." *Journal of Political Economy* 86 (2):S87–S120.
Ibarra, Gabriel Lara, David McKenzie, and Claudia Ruiz Ortega. 2017. Learning the Impact of Financial Education When Take-Up Is Low. *Policy Research Working Paper no. 8238*. Washington DC: World Bank.
IED (Independent Evaluation Department). 2013. Thematic Evaluation Study on ADB's Support for Achieving the Millennium Goals. Manila: ADB.
IED (Independent Evaluation Department). 2016. Mitigating the Impacts of Climate Change and Natural Disasters for Better Quality Growth. Manila: ADB.
IMF (International Monetary Fund). 2015. Counting the Cost of Energy Subsidies. Washington DC: IMF.
IMF Staff. 2001. Global Trade Liberalization and the Developing Countries. International Monetary Fund. Washington DC: IMF.
IPCC (Intergovernmental Panel on Climate Change). 2018. The Summary for Policymakers of the Special Report on Global Warming of 1.5°C (SR15). IPCC.
Jiang, Shiqing, Ming Lu, and Hiroshi Sato. 2012. "Identity, Inequality, and Happiness: Evidence from Urban China." *World Development* 40 (6):1190–1200.
Juric, Radmilla, Inhwa Kim, Hemalata Paneerselvam, and Igor Tesanovic. 2017. Analysis of ZIKA Virus Tweets: Could Hadoop Platform Help in Global Health Management. *Proceedings of the 50th Hawaii International Conference on System Sciences*.
Kahneman, Daniel, Peter Wakker, and Rakesh Sarin. 1997. "Back to Bentham? Explorations of Experienced Utility." *Quarterly Journal of Economics* 112 (2):375–406.
Kanbur, Ravi. 2017. What is the World Bank Good For? Global Public Goods and Global Institutions. *CEPR Discussion Paper no. DP12090*.
Kaufmann, Daniel, Aart Kraay, and Massimo Mastruzzi. 2010. The Worldwide Governance Indicators: Methodology and Analytical Issues. *Policy Research Working Paper no. 5430*.

Kremer, Michael, and Christopher M. Snyder. 2015. "Preventives Versus Treatments." *Quarterly Journal of Economics* 130 (3):1167–1239.

Li, Zhengtao, Henk Folmer, and Jianhong Xue. 2014. "To What Extent Does Air Pollution Affect Happiness? The Case of the Jinchuan Mining Area, China." *Ecological Economics* 99:88–99.

Mohler, G. O., M. B. Short, S. Malinowski, M. Johnson, G. E. Tita, A. L. Bertozzi, and P. J. Brantingham. 2015. "Randomized Controlled Field Trials of Predictive Policing." *Journal of the American Statistical Association* 110 (512):1399–1411.

Ostry, Jonathan, Andrew Berg, and Charalambos Tsangarides. 2014. Redistribution, Inequality and Growth. *IMF Staff Discussion Note*. Washington DC: IMF.

Petri, Peter, and Vinod Thomas. 2013. Development Imperatives for the Asian Century. *ADB Economics Working Paper Series no. 360*. Manila: ADB.

Rehdanz, Katrin, and David Maddison. 2005. "Climate and Happiness." *Ecological Economics* 52 (1):111–125.

Roemer, John. 1993. "A Pragmatic Theory of Responsibility for the Egalitarian Planner." *Philosophy and Public Affairs* 22 (2):146–166.

Sachs, Jeffrey D. 2012. "From Millennium Development Goals to Sustainable Development Goals." *Lancet* 379 (9832):2206–2211.

Sachs, Jeffrey D. 2015. *The Age of Sustainable Development*. New York: Columbia University Press.

Sommer, Alfred. 2016. "Burning Fossil Fuels: Impact of Climate Change on Health." *International Journal of Health Services* 46 (1):48–52.

Thomas, Vinod. 2018. *Climate Change and Natural Disasters: Transforming Economies and Policies for a Sustainable Future*: Routledge. Taylor & Francis Group.

Thomas, Vinod, Dailami Mansoor, Ashok Dhareshwar, Daniel Kaufmann, Nalin Kishor, Ramon Lopez, and Yan Wang. 2000. *The Quality of Growth*. New York: Oxford University Press.

UN Global Pulse. 2011. Monitoring Perceptions of Crisis-Related Stress Using Social Media Data. Accessed 11 November, 2017.

UN Global Pulse. 2012. Big Data for Development: Challenges & Opportunities. UN Global Pulse.

UN Global Pulse. 2014. Mining Indonesian Tweets to Understand Food Price Crises. Jakarta: UN Global Pulse.

US GAO (Government Accountability Office). 2017. CMS Fraud Prevention System Uses Claims Analysis to Address Fraud GAO-17-710. Washington, DC: US GAO.

Van den Berg, Rob D. 2011. "Evaluation in the Context of Global Public Goods." *Evaluation* 17 (4):405–417.

Van den Berg, Rob D., Indran Naidoo, and Susan D. Tamondong, eds. 2018. *Evaluation for Agenda 2030. IDEAS, UNDP*. Exeter, United Kingdom: International Development Evaluation Association.

WWF (World Wildlife Fund). 2018. Wildlife Declines Show Nature Needs Life Support. *Living Planet Report 2018*.

Zie, Julie. 2015. "Reporting Potholes: There Are Too Many Apps for That." http://www.boston.com/cars/news-and-reviews/2015/01/22/reporting-potholes-there-are-too-many-apps-for-that/fffZvt3LME056eBbSmDtEM/story.html.

Zivin, Graff Joshua, and Matthew Neidell. 2012. "The Impact of Pollution on Worker Productivity." *American Economic Review* 102 (7):3652–3673.

Open Access This chapter is licensed under the terms of the Creative Commons Attribution 4.0 International License (http://creativecommons.org/licenses/by/4.0/), which permits use, sharing, adaptation, distribution and reproduction in any medium or format, as long as you give appropriate credit to the original author(s) and the source, provide a link to the Creative Commons licence and indicate if changes were made.

The images or other third party material in this chapter are included in the chapter's Creative Commons licence, unless indicated otherwise in a credit line to the material. If material is not included in the chapter's Creative Commons licence and your intended use is not permitted by statutory regulation or exceeds the permitted use, you will need to obtain permission directly from the copyright holder.

Correction: Economic Evaluation of Sustainable Development

Vinod Thomas and Namrata Chindarkar

The original version of this chapter was revised.

Correction to:

Chapter 1 in: Vinod Thomas and Namrata Chindarkar, Economic Evaluation of Sustainable Development DOI 10.1007/978-981-13-6389-4_1

In Figure 1.1, the word 'governance' has been added in the middle of the image.

Fig. 1.1 Interactions among sustainable development issues. (Source: Authors' illustration)

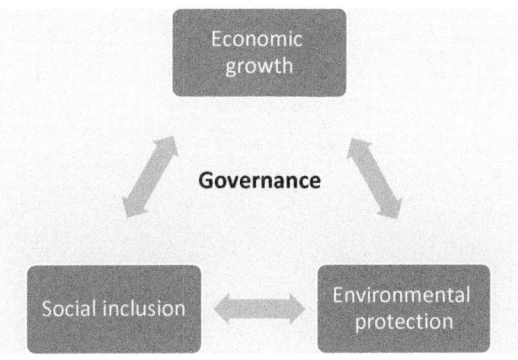

The updated original online version for this chapter can be found at https://doi.org/10.1007/978-981-13-6389-4_1

Index

A
Attrition, 37
Average treatment effect (ATE), 33
Average treatment effect on the treated (ATT), 35

B
Baseline surveys, 36
Before-and-after comparison, 29
Big data, 134–137
Brazil, 26

C
Causal inference, 27–28
Causality, 27
China, 41
Common support, 49
Consumer surplus, 75
Contingency allowance, 70
Contingent valuation, 71
Control group, 29
Cost-benefit analysis (CBA), 2, 11, 64–65
Cost-effectiveness analysis, 85
Counterfactuals, 27–31

D
Deadweight loss, 75
Deforestation, 16, 26
Development impact, 99
Difference-in-differences (DID), 37–41
Discount rate, 77
Distributional weights, 82
Double counting, 73–74

E
Effectiveness, 98
Efficiency, 98
Electrification upgrading program, 27
Eligibility index, 41
Endogeneity, 44

Environmental protection, 15–16, 130
Environmental stewardship, 4
Exclusion restriction, 46
Exogeneity, 45
Experimental, 36
Externalities, 34, 75–76
External validity, 32–37

F
Forest strategy, 117

G
Global public goods (GPGs), 132–134
Governance, 4, 16–18, 131

H
Hedonic pricing, 71

I
Impact evaluation (IE), 2, 11, 26–31
Inclusion, 128–129
Inclusive growth, 13–15
India, 28, 88–90, 112
Indirect benefits, 69
Indirect costs, 71
Inequality, 13
Instrumental variables (IV), 44–48
Intent to treat (ITT), 36
Internal validity, 32–37

L
Law of large numbers (LLN), 32
Local average treatment effect (LATE), 36

M
Marginal social cost, 76
Measurement error, 32
Mexico, 85–88
Millennium Development Goals (MDGs), 7
Mozambique, 66
Multilateral development, 12

N
Net present value (NPV), 77–84
Net social benefits, 74–76

O
Objectives-based evaluation (OBE), 96–100
Omitted variables, 32
Opportunity cost, 68
Overseas development assistance (ODA), 96

P
Parallel-trends, 40
Pareto improvement, 81
The Philippines, 48
Piketty, Thomas, 13
Pollution, 43
Probit regression, 49
Producer surplus, 75
Project performance assessment report (PPAR), 103
Propensity score matching (PSM), 48–52

Q
Quasi-experimental, 37–52

R

Random assignment, 32–33
Randomized controlled trials (RCTs), 32
Redistribution, 13
Regression discontinuity design (RDD), 41–44
Relevance, 98
Rural electrification, 66

S

Selection bias, 31
Selection mechanism, 31
Sensitivity analysis, 80
Shadow price, 70
Social discount rate, 78
Social inclusion, 4
Spillovers, 27
Sri Lanka, 102
Stern Review, 79
Sustainability, 99
Sustainable Development Goals (SDGs), 7–8

T

Theory of change, 31
Time-declining discount rate, 78
Travel-cost technique, 71
Treatment effect, 27
Treatment group, 29
Two-stage least-squares (2SLS), 46

U

Unconfoundedness assumption, 50
The United States, 45

V

Valuation, 70

W

Willingness-to-pay (WTP), 70
With-and-without comparison, 30

The manufacturer's authorised representative in the EU is Springer Nature Customer Service Centre GmbH, Europaplatz 3, 69115 Heidelberg, Germany. If you have any concerns regarding our products, please contact ProductSafety@springernature.com

Printed and bound by CPI Group (UK) Ltd, Croydon, CR0 4YY

25/03/2026

02078175-0016